Ronny Henker

Brillouin Slow- and Fast-Light

Ronny Henker

Brillouin Slow- and Fast-Light

Theory and Experiments

Südwestdeutscher Verlag für
Hochschulschriften

Imprint
Any brand names and product names mentioned in this book are subject to trademark, brand or patent protection and are trademarks or registered trademarks of their respective holders. The use of brand names, product names, common names, trade names, product descriptions etc. even without a particular marking in this work is in no way to be construed to mean that such names may be regarded as unrestricted in respect of trademark and brand protection legislation and could thus be used by anyone.

Publisher:
Südwestdeutscher Verlag für Hochschulschriften
is a trademark of
Dodo Books Indian Ocean Ltd., member of the OmniScriptum S.R.L Publishing group
str. A.Russo 15, of. 61, Chisinau-2068, Republic of Moldova Europe
Printed at: see last page
ISBN: 978-3-8381-2039-3

Zugl. / Approved by: Dublin, Dublin Institute of Technology, Diss., 2010

Copyright © Ronny Henker
Copyright © 2010 Dodo Books Indian Ocean Ltd., member of the OmniScriptum S.R.L Publishing group

Abstract

In today's information age demand for ultra-fast information transfer with ultra-high bandwidths has reached extraordinary levels. Hence, the transmission in the future internet-backbone will be increasingly constrained in the network nodes. At the same time, the power consumption of the network systems will increase to unsustainable levels. To overcome these constraints power-efficient photonic networks which provide ultra-fast all-optical switching and routing are essential. Nowadays, optical signal processing and switching can be implemented relatively easily. However, the realization of optical buffers and short-term memories is still an unsolved challenge.

The slow- and fast-light effect has been investigated as one solution for the optical buffering over the last few years. It means the slowing down and acceleration of the group velocity of light pulses in a medium. To realize this, many different methods and material systems have been developed but due to its significant advantages the nonlinear effect of stimulated Brillouin scattering (SBS) is particularly promising. However, it also suffers from disadvantages which limit the slow- and fast-light performance.

In this work the slow- and fast-light effect using SBS is investigated. SBS-based slow- and fast-light utilizes the generation of artificial resonances by the coupling of a strong pump wave inside an optical fiber. This creates gain and loss regions in which a counterpropagating signal wave can be amplified and attenuated, but slowed down and accelerated as well. The properties of such a system depend directly on the Brillouin spectrum. Thus, the focus of this work is the analysis of opportunities to overcome the natural limits of SBS-based slow-light by the optimization of the transfer function. Therefore, different novel methods for a bandwidth enhancement, an enhancement of the maximum achievable time delay and a reduction of the output pulse distortions are investigated in theory and experiment. This includes the theoretical background and practical limits of these methods. The most important results are a bandwidth enhancement to a multiple of the natural Brillouin bandwidth up to the GHz range, a time delay enhancement of up to four times of the initial pulse width and a significant distortion reduction of the output pulses by 25 %.

Although there are various potential applications for the slow- and fast-light effect, this work examines the topic mainly under the aspect of information and communication systems. Thus, this work will show that SBS-based slow- and fast-light is a reliable and efficient candidate to realize all-optical short-term buffers. Despite the small storage capacity compared to other techniques, it can be used for an accurate synchronization, multiplexing and equalization of multiple high bit rate data channels as well as for phased-array antennas.

Acknowledgements

This work was carried out within the framework of my doctoral studies at the School of Electronic and Communications Engineering of Dublin Institute of Technology, Ireland. First of all, I wish to express my sincere gratitude to Prof. Dr. Thomas Schneider of the Hochschule für Telekommunikation Leipzig for his encouragement and inspiration for the research on this topic. His guidance and advice was invaluable throughout time. Furthermore, I am also indebted to Dr. Max Ammann, Dr. Andreas Schwarzbacher and Dr. Gerald Farrell from the School of Electronic and Communications Engineering at the Dublin Institute of Technology for the opportunity to work on the Ph.D. and for all their support.

A very special thanks go to all of my colleagues for all their help and support that I have received throughout the time. They have provided an essential part to the work by fruitful discussions, theoretical and technical aid. In particular I would like to thank Markus Junker and Kai-Uwe Lauterbach for the help with the technical equipment, Andrzej Wiatrek for the assistance with the simulations and Jens Klinger for everything concerned with the laboratory. Furthermore, I wish Andrzej Wiatrek, Stefan Preußler and Kambiz Jamshidi all the best in continuing the work laid out in this book.

I would also like to acknowledge the organizational support of the academic administration of the Hochschule für Telekommunikation Leipzig and the Dublin Institute of Technology. Additionally, I want to gratefully acknowledge the research groups of Prof. Dr.-Ing. Schäffer at TU Dresden and of Prof. Dr. Buse at University Bonn for the loan of technical equipment which was essential for the realization of different measurements.

Finally, and most of all, I would like to express my deep gratitude to my parents and friends for their moral support and understanding over the last few years. A particular thank you goes especially to Kristin for her patience, sacrifice, understanding and her belief in me.

I would also like to thank the Deutsche Telekom AG for the financial support.

Table of Contents

Abstract — I

Acknowledgements — III

List of Abbreviations and Symbols — IX

List of Figures — XV

List of Tables — XIX

1 Introduction — 1
   1.1 Slow- and Fast-Light — 1
   1.2 Objectives and Outline of the Book — 3

2 Fundamentals — 5
   2.1 Velocities of Light — 5
      2.1.1 Photon Velocity — 5
      2.1.2 Phase Velocity — 5
      2.1.3 Group Velocity — 6
      2.1.4 Signal or Information Velocity — 9
   2.2 Refractive Index and Kramers-Kronig Relations — 11
   2.3 Figures of Merit and Metrics — 12
      2.3.1 Group Index, Group Velocity, Group Delay, Time Delay, Slowdown Factor — 13
      2.3.2 Fractional and Effective Time Delay — 13
      2.3.3 Broadening Factor — 14
      2.3.4 Delay-Bandwidth Product, Delay-Bit Rate Product — 14
      2.3.5 Q-Delay Product — 15
      2.3.6 Other Metrics — 15
   2.4 Slow- and Fast-Light Methods — 16
      2.4.1 Waveguide-Loops — 18
      2.4.2 Electromagnetically Induced Transparency and Coherent Population Oscillations — 18
      2.4.3 Engineered Materials and Structures — 21
      2.4.4 Optical Amplifiers — 22
      2.4.5 Optical Fibers — 23

2.4.6 Wavelength-Conversion and Dispersion 24
2.4.7 Quasi-Light Storage 26
2.5 Applications of Slow- and Fast-Light 26
2.5.1 Optical Buffers and Memories in Packed-Switched Networks 27
2.5.2 Spectroscopy and Interferometry 29
2.5.3 Nonlinear Processes 30
2.5.4 Phased-Array Antennas 30
2.5.5 Quantum Information Processing 31
2.6 Summary 31

3 Slow- and Fast-Light Based on Stimulated Brillouin Scattering 33
3.1 Brillouin Scattering 33
3.1.1 Generation and Mode of Operation 34
3.1.2 Brillouin Threshold and Brillouin Gain Coefficient 35
3.1.3 Intensity Equations and Brillouin Amplifier Gain 38
3.1.4 Spectral Characteristics 41
3.2 SBS-Based Slow- and Fast-Light 43
3.2.1 Background Theory 43
3.2.2 Principle Experimental Setup 49
3.2.3 Preliminary Basic Investigations and Measurements 50
3.3 Properties of SBS-Based Slow- and Fast-Light 54
3.4 Summary 55

4 Brillouin Bandwidth Enhancement 57
4.1 Introduction 57
4.2 Experimental Verification and Results 59
4.2.1 Polychromatic Pump Sources 59
4.2.2 Broadband Pump Sources 61
4.3 Summary 64

5 Time Delay Enhancement 65
5.1 Introduction 65
5.2 Superposition of a Narrow Brillouin Gain With a Broadband Brillouin Loss . . . 67
5.2.1 Background Theory 67
5.2.2 Experimental Verification and Results 70
5.3 Superposition of a Brillouin Gain With Two Brillouin Losses 74
5.3.1 Background Theory 74
5.3.2 Optimization 79
5.3.3 Experimental Verification and Results 82

	5.4	Suppression of Spurious Backscattered Stokes-Waves in Multiple-Pump-Line Systems 86
		5.4.1 Background Theory 86
		5.4.2 Experimental Verification and Results 89
	5.5	Summary 92

6 Pulse Distortion Reduction   95

     6.1 Introduction 95
     6.2 Background Theory 97
     6.3 Experimental Verification and Results 103
     6.4 Summary 106

7 Fast-Light   107

     7.1 Introduction 107
     7.2 Background Theory 108
     7.3 Experimnental Verification and Results 110
     7.4 Summary 112

8 Limitations of SBS-Based Slow- and Fast-Light   113

     8.1 Introduction 113
     8.2 Limitations of the Investigated Systems 114
         8.2.1 Maximum Gain, Maximum Time Delay and Storage Capacity 115
         8.2.2 Pump Power Degradation by Four-Wave Mixing 117
     8.3 Summary 119

9 Conclusions   121

10 Future Work   125

References   127

A Measurement of Brillouin Parameters   149

     A.1 Brillouin Shift Measurement 149
     A.2 Brillouin Threshold Measurement 151
     A.3 Brillouin Spectrum and Bandwidth Measurement 154
     A.4 Summary Fiber Parameters 156

B Pulse Diagrams   159

     B.1 Single Natural Brillouin Gain 159
     B.2 Single Narrow Brillouin Gain Superimposed With a Broad Brillouin Loss 161
     B.3 Single Natural Brillouin Gain Superimposed With Two Narrow Brillouin Losses   162
     B.4 Single Broadened Brillouin Gain Superimposed With Two Narrow Brillouin Losses 163

B.5 Suppression of Spurious Backscattered Stokes-Waves via FBG and Filters 163
B.6 Distortion Reduction in Cascaded Slow-Light Systems 165

Index **167**

List of Abbreviations and Symbols

Abbreviations

ASE	Amplified spontaneous emission
BER	Bit error ratio
$BPSK$	Binary phase-shift-keying
C	Optical circulator
CPO	Coherent population oscillations
$CROW$	Coupled resonator optical waveguide
CW	Continuous wave
DBP	Delay-bandwidth product, delay-bit-rate product
DFB	Distributed feedback
$DPSK$	Differential phase-shift-keying
DSF	Dispersion-shifted fiber
$EDFA$	Erbium-doped fiber amplifier
EIT	Electromagnetically induced transparency
ESA	Electrical spectrum analyzer
FBG	Fiber Bragg grating
FUT	Fiber under test
$FWHM$	Full width at half maximum
FWM	Four-wave mixing
GVD	Group velocity dispersion
$HNLF$	Highly-nonlinear fiber
$HWHM$	Half width at half maximum
ISO	Optical Isolator
KKR	Kramers-Kronig relations
LD	Laser diode
MZM	Mach-Zehnder modulator
NRZ	Non-return-to-zero
OEO	Optical-electrical-optical
OPC	Optical polarization controller
OPM	Optical power meter
OSA	Optical spectrum analyzer
PC	Photonic crystal
PD	Photodiode
PM	Phase modulator

$PPLN$	Periodically poled lithium niobate
$PRBS$	Pseudo-random binary sequence
QLS	Quasi-light storage
SBS	Stimulated Brillouin scattering
SOA	Semiconductor optical amplifier
SOI	Silicon-on-insulator
SRS	Stimulated Raman scattering
$SSMF$	Standard single mode fiber
TF	Tunable filter
$TFBG$	Tunable fiber Bragg grating
WDM	Wavelength division multiplexing

Symbols

α	Absorption coefficient, attenuation
α_a	Acoustic attenuation constant
$\alpha_{dB/km}$	Attenuation constant in [dB/km]
$\alpha_{km^{-1}}$	Attenuation constant in [km^{-1}]
β	Nonlinear coefficient
χ	Susceptibility
χ'	Real part of susceptibility
χ''	Imaginary part of susceptibility
Δf	FWHM bandwidth, bit rate
Δf_B	Brillouin FWHM bandwidth
Δf_p	FWHM linewidth of the pump source
Δk	Variation of wave number, phase mismatch
Δk_I	Imaginary part of complex phase mismatch term
Δk_R	Real part of complex phase mismatch term
Δn_g	Group index change
ΔT	Absolute time delay
ΔT_{eff}	Effective time delay
ΔT_{eff}^{max}	Maximum effective time delay
ΔT_{frac}	Fractional time delay
ΔT_{frac}^{max}	Maximum fractional time delay
δ	Angular frequency separation of the loss from the line center ω_0
$\Delta\lambda$	Wavelength shift
$\Delta\omega$	Variation of angular frequency, angular FWHM bandwidth
$\Delta\omega_B$	Angular Brillouin FWHM bandwidth
$\Delta\omega_P$	Angular pulse FWHM bandwidth
η	Mixing efficiency for FWM

List of Abbreviations and Symbols

γ	Angular HWHM bandwidth
γ_G	Angular half $1/e$-bandwidth
$\hat{\varepsilon}_r$	Complex relative permittivity
$\hat{n}$	Complex refractive index
$\Im$	Imaginary part
λ	Wavelength
λ_c	Carrier wavelength
λ_{in}	Input wavelength
λ_{out}	Output wavelength
λ_p	Pump wavelength
μ_r	Relative magnetic permeability
Ω	Normalized angular frequency
ω	Angular frequency
ω_0	Angular central frequency, angular line center frequency
$\otimes$	Convolution operation
Φ	Phase
$\Re$	Real part
τ_{in}	Input FWHM pulse width
τ_{out}	Output FWHM pulse width
$erfc$	Complementary error function
ε_r	Relative permittivity
A_{eff}	Effective area
B	Broadening factor
B_{GVD}	GVD-dependent broadening factor
c	Speed of light in a vacuum
$c.c.$	Conjugate-complex
D	Linear insertion loss
d	Normalized frequency separation
D_{GVD}	GVD parameter
DBP	Delay-bandwidth product, delay-bit-rate product
E_p	Pump field
E_s	Signal or Stokes-field
E_0	Wave amplitude
f	Frequency
f_B	Brillouin frequency shift
f_a	Acoustic frequency
f_{mod}	Modulation frequency
f_p	Pump frequency
f_s	Stokes frequency, signal frequency
G	Linear Brillouin amplification factor

g	Brillouin gain
g_B	Polarization-dependent Brillouin gain coefficient
$g_B^{eff}(f)$	Effective/final Brillouin spectrum
g_{Bmax}	Maximum Brillouin gain coefficient without polarization influence
G_{dB}	Logarithmic Brillouin amplification factor
G_{dB}^{th}	Logarithmic Brillouin gain threshold
g_{max}	Maximum Brillouin amplification gain
g_{td}	Time-delay-gain
g_{th}	Linear Brillouin gain threshold
$H(\omega)$	Transfer function in the frequency domain
i	Index, number of pass
I_p	Pump intensity
I_s	Stokes-intensity
k	Loss-gain-bandwidth ratio
$k(\omega)$	Complex wave number as a function of the frequency
k_1	Reciprocal group velocity
k_2	Group velocity dispersion
k_0	Wave number in a vacuum
K_B	Polarization factor between pump and Stokes-wave
$k_{SBS}(\omega)$	Complex Brillouin-based wave number as a function of the frequency
L	Length
L_{eff}	Effective length
m	Loss-gain-ratio
M_2	Elasto-optic figure of merit
N	Number of bits
n	Refractive index
n'	Real part of refractive index
n''	Imaginary part of refractive index
n_g	Group index
P_p	Pump power
$P_p(f)$	Pump spectrum
P_s	Stokes power
P_p^{th}	Brillouin power threshold
Q	Q-factor
S	Slowdown factor
t	Time
T_a	Lifetime of acoustic phonons
t_g	Group delay
t_{del}	Pulse delay time
t_{ph}	Phase delay

v_a	Speed of sound
v_g	Group velocity
v_{ph}	Phase velocity
z	Distance

List of Figures

2.1 Group velocity and pulse delay as a function of $\omega_0 dn(\omega)/d\omega$ 8
2.2 Absorption coefficient and refractive index as a function of the frequency detuning normalized to the absorption FWHM bandwidth 12
2.3 Classification of the slow- and fast-light methods according to the basic slow- and fast-light technique 16
2.4 Classification of the slow- and fast-light methods according to the nature of the pulse propagation time alteration 17
2.5 Absorption coefficient and refractive index as a function of the frequency detuning for an EIT scheme 20
2.6 Applications of the slow- and fast-light effect 27
2.7 Packet contention resolution in a switch 28

3.1 Generation of SBS 35
3.2 Calculated and measured Brillouin threshold 36
3.3 Propagation conditions of the pump and Stokes-wave along the fiber 39
3.4 Calculated and measured Brillouin gain as a function of the fiber length for two constant pump powers 40
3.5 Measured Brillouin gain spectrum in comparison to a Lorentzian and Gaussian distribution 42
3.6 Stokes-gain and anti-Stokes-loss, phase distribution, and group index change due to SBS 43
3.7 Behavior of a Brillouin gain resonance 47
3.8 Principle experimental setup of Brillouin-based slow- and fast-light systems . . . 49
3.9 Time delay slope and pump power requirements for a SBS-based slow-light system with a single natural Brillouin gain 51
3.10 Principle experimental setup of a SBS-based slow-light system with a single natural Brillouin gain 52
3.11 Measured time delay and time functions of the delayed pulses 53
3.12 Pulse broadening as a function of the pump power in a SBS-based slow-light system with a single natural Brillouin gain and variable SSMF length 54
3.13 Saturation of the time delay in a SSMF with a length of 25 km 55

4.1 Convolution of the pump spectrum with the natural Brillouin gain spectrum . . 58

4.2 Experimental pump system for Brillouin bandwidth enhancement via external modulation of the pump source 60
4.3 Brillouin bandwidth enhancement via multiple discrete pump lines 60
4.4 Experimental pump system for Brillouin bandwidth enhancement via direct modulation of the pump source 61
4.5 Brillouin bandwidth enhancement via direct modulation of one pump source .. 61
4.6 Idea of Brillouin bandwidth enhancement via direct modulation of two pump sources 62
4.7 Practical Brillouin bandwidth enhancement via direct modulation of two pump sources 63

5.1 Behavior of a superposition of a gain with a broadened loss in comparison to a natural Brillouin gain 68
5.2 Experimental verification for a time delay enhancement via superposition of a narrow Brillouin gain with a broadened Brillouin loss 70
5.3 Measurement results for the proof of concept of a time delay enhancement via superposition of a narrow Brillouin gain with a broadened Brillouin loss 71
5.4 Time delay enhancement and zero-gain slow-light via superposition of a narrow Brillouin gain with a broadened Brillouin loss 73
5.5 Behavior of a superposition of a natural gain with two losses at its spectral boundaries 75
5.6 Behavior of a superposition of a broadened gain with two losses at its spectral boundaries in comparison to a single normal and a broadened gain 79
5.7 Normalized line center group index change as a function of the normalized frequency separation of the loss spectra for different relations between the gain and the losses with $\gamma_1 = \gamma_2$ 80
5.8 Optimization of the normalized group index change, line center time delay and FWHM bandwidth by the variation of the loss separation for a superposition of a Brillouin gain with two losses at its spectral boundaries 81
5.9 Experimental verification for a time delay enhancement via superposition of a Brillouin gain with two Brillouin losses 82
5.10 Optimization of the normalized time delay, output pulse width and effective time delay as a function of the loss separation for the superposition of a Brillouin gain with two Brillouin losses at its spectral boundaries 83
5.11 Measurement results for the proof of concept of a time delay enhancement via superposition of a Brillouin gain with two losses at its spectral boundaries ... 84
5.12 Measurement results for a time delay enhancement via superposition of a natural and broadened Brillouin gain with two Brillouin losses at its spectral boundaries 85
5.13 Overall gain and time delay due to a superposition of a Brillouin gain with two Brillouin losses at its spectral boundaries 87

List of Figures

5.14 Suppression of spurious backscattered Stokes-waves and compensation of the insertion loss of the filter stages 88
5.15 Slow-light medium configurations for spurious Stokes-wave suppression 89
5.16 Measurement results for time delay enhancement via suppression of backscattered spurious Stokes-waves 91

6.1 Temporal broadening of 120 ns delayed pulse 95
6.2 Pulse distortion reduction via Brillouin bandwidth broadening 98
6.3 Normalized time delay as a function of bandwidth ratio k and the normalized frequency separation of the losses d for a superposition of a Brillouin gain with two losses with a loss-gain-ratio $m = 1$ 99
6.4 Pulse distortion reduction by Brillouin bandwidth broadening via multiple discrete pump lines. 100
6.5 Normalized gain and line center group index change for a pure Gaussian-shaped Brillouin spectrum and a Gaussian-shaped Brillouin spectrum superimposed with two losses at its spectral boundaries 101
6.6 Normalized group index change and GVD for a natural gain, a broadened gain and a broadened gain superimposed with two losses at its spectral boundaries . 101
6.7 Behavior of a superposition of a doubled Brillouin gain with two losses at its spectral boundaries in comparison to a single gain 102
6.8 Experimental verification of pulse distortion reduction in cascaded slow-light systems via external pump modulation 103
6.9 Measurement results for the distortion reduction via superposition of a Brillouin gain with two losses at its spectral boundaries 105

7.1 Behavior of a superposition of a narrow loss with a broadened gain 109
7.2 Results of fast-light measurements 111

8.1 Maximum achievable time-delay-gain and time delay as a function of the input pulse power for a SBS-based slow-light system with a single gain, a narrow gain superimposed with a broad loss and a gain superimposed with two losses at its spectral boundaries in a SSMF with a length of 50 km 116
8.2 Pump power degradation by FWM 118

A.1 Measurement setups for the Brillouin shift 149
A.2 Output spectra of Brillouin shift measurement 150
A.3 Measurement setup for Brillouin threshold 151
A.4 Brillouin threshold measurement of SSMF with different lengths *(to be continued)* 152
A.4 Brillouin threshold measurement of SSMF with different lengths *(continued)* . . 153
A.5 Measurement setup for Brillouin spectrum 154
A.6 Brillouin spectrum measurement of a SSMF with a length of 5 km 155

B.1 Delayed pulses for a single natural Brillouin gain *(to be continued)* 159
B.1 Delayed pulses for a single natural Brillouin gain *(continued)* 160
B.2 Delayed pulses for a superposition of a single natural Brillouin gain with a broad Brillouin loss 161
B.3 Delayed pulses for a superposition of a natural Brillouin gain with two Brillouin losses at its spectral boundaries 162
B.4 Delayed pulses for a superposition of a broadened Brillouin gain with two Brillouin losses at its spectral boundaries 163
B.5 Delayed pulses for a blocking via two segments with one FBG 163
B.6 Delayed pulses for a blocking via three segments with two FBG 164
B.7 Delayed pulses for a blocking via two segments with one WDM filter 164
B.8 Delayed pulses for a distortion reduction via external pump modulation in one segment of cascaded slow-light systems 165

List of Tables

A.1 Constants of all used fibers 156
A.2 Different parameters of the used fibers 157

1 Introduction

The physics of light propagation has a long exciting history. The first successful estimation of the speed of light was carried out by Rømer and Huygens via astronomical time delay observations in 1676 and 1678 [1]. In 1849 Fizeau determined the terrestrial light velocity for the first time with a toothed wheel method, although he overstated the value by 5 % [2]. The group velocity was then fully described by Lord Rayleigh in 1877 [3]. Even the studies of light propagation in media with reduced group velocities dates back to the nineteenth century when Lorentz and others worked on the classical theory of the dispersion of electromagnetic waves [4]. During the twentieth century the estimation of superluminal group velocities was a very intriguing research topic due to the conflict with Einstein's theory of special relativity and causality, which states that no signal can propagate faster than the speed of light in a vacuum [1, 5]. With the work focused on signal, group and energy propagation velocity by Sommerfeld and Brillouin this curio could be clarified [6, 7].

Nowadays at the beginning of the twenty-first century, the phenomenon of speed of light is still a very interesting topic. For few years there has been an increasing number of researchers who are working on reduced and accelerated speed of light in media for which the terms *slow- and fast-light* had been established. Besides the fundamental physical interest this approach has very high practical potential. The increasing demand on higher bandwidths for future Internet services requires the development of all-optical networks with ultrahigh speed photonic switching and routing. At the same time, such an all-optical network can drastically reduce the energy requirements and therefore, the carbon footprint of communication and information systems [8, 9]. Another problem of todays networks is that in most cases the signals are converted from the optical domain into the electrical where they are proceeded and then converted back to optical domain. For increasing carrier frequencies and increasing signal bandwidths this would not be possible without suffering from significant losses. If the signal processing is completely optical this is not a problem because every optical fiber is able to transmit more than 100 channels with data rates of tens of Gbit/s over large distances.

1.1 Slow- and Fast-Light

While most functions of such an all-optical network have already been shown [10, 11], the indispensable task to buffer optical signals has to be solved in the near future. The slow-light effect is a very promising tool to implement this and can be seen as a key technology for all-optical networks. It provides not only the technique for optical buffers and memories, but also enables a way to design for example optical synchronizers, equalizers and signal processors. Among the use in optical communication and information systems, there are many more applications possible. Slow-light can be used for instance for time-resolved spectroscopy, microwave photonics and nonlinear optics [12]. Due to this very high practical potential a large interest on slowing down or acceleration of light has been arisen over the last few years. Therefore, the United States of America and the European Union fund separate research programs on the development of effective technologies for enabling slow- and fast-light propagation (DARPA Slow-Light and GOSPEL Project). Furthermore, the Optical Society of America organizes a periodic conference exclusively on this topic.

Many different methods and material systems have been proposed for creating slow- and fast-light: from atomic-gases via semiconductors through to optical amplifiers and fibers. In principle, all of them can be divided into three different categories:

1. methods which rely on the physical length of a waveguide or resonator,

2. methods which are based on a change of the velocity of propagation and

3. methods on the basis of the time-frequency-coherence of signals.

In this book only the second category is investigated. The propagation velocity of the pulses can be changed by creating artificial tailored resonances or dispersions inside the material. Hence, a continuous alteration of the group velocity of light pulses is provided. Therefore, the time delay in the material with constant length can be controlled externally and variably within a particular range. However, one problem of this method is that the delay or acceleration of the pulses is accompanied by distortions of the pulse shape.

A very promising opportunity to create and control an artificial dispersion inside an optical fiber is the nonlinear effect of stimulated Brillouin scattering (SBS) [13, 14]. On the one hand, SBS offers several advantages over the other methods and the group velocity can be controlled over a very wide range [15]. Just small pump powers are necessary to achieve very high time delays. The systems are very easy to implement and can be built using standard components of telecommunications [16]. Furthermore, the SBS works in all fiber types in their entire transparency range which makes the systems very flexible. Last but not least, there is the great advantage that the fibers themselves are used as slow- and fast-light medium and hence, such schemes can be integrated into existing optical systems easily and seamlessly.

On the other hand, the SBS introduces some disadvantages which limits the performance of the slow- and fast-light systems. The maximum delayable data rate is restricted by the narrow natural full width at half maximum (FWHM) Brillouin bandwidth of around 30 MHz in standard single mode fibers (SSMF) at wavelengths of 1550 nm [17]. Furthermore, the maximum achievable time delay is limited by the saturation of the Brillouin amplifier. Another problem is that the time delay of the pulses is accompanied by a distortion of the pulse shape which primarily manifests itself in a temporal broadening of the pulse width.

1.2 Objectives and Outline of the Book

The main objective of this work is to analyze and investigate the slowing down (slow-light) and acceleration (fast-light) of optical pulses in optical fibers via the nonlinear effect of SBS as preliminary inquiry. This book covers the fundamentals and natural limits of SBS-based slow- and fast-light systems. Therefore, only optical single pulses are used to determine the properties of the effect. All measurements were carried out in the third optical window at wavelengths around 1550 nm. Furthermore, the investigations are accomplished using only SSMF as slow- and fast-light medium because this fiber type is the most prevalent in the optical backbone network. Hence, prospective systems could be easily included into the existing communication network. Due to the disadvantages of the SBS, as mentioned in the last section, this work primarily focuses on three topics: the enhancement of the Brillouin bandwidth, the enhancement of the SBS-induced time delay and the reduction of the SBS-induced pulse distortions. Therefore, different SBS schemes are presented to achieve high time delays, wide bandwidths and less distortions. Based on these schemes, it will be examined and evaluated how far the effect is applicable within optical communication and information systems. This is the basis for further investigations with real data streams and the future implementation into existing systems.

This book is divided into 10 chapters. After this introduction the following chapter describes the fundamentals of slow- and fast-light. At first, the different velocities of light are defined and clarified. This includes the photon, phase, group and signal velocity. As will be shown, there are significant differences between these velocities and therefore, the dispute with the theory of special relativity regarding superluminal signals is resolved. Furthermore, the Kramers-Kronig relations (KKR) are briefly introduced as a mathematical physical tool which relates the real and imaginary part of frequency-dependent quantities [18]. With the aid of this tool it is possible to explain the relation between the absorption/gain of the SBS and the dispersion and therefore, the refractive index change, respectively. In the third section of this chapter the metrics of slow- and fast-light are defined. They are necessary to evaluate the grade of delay and the efficiency of the systems. Special attention in this chapter is given to the possible applications and generation methods of the slow- and fast-light effect.

In Chapter 3 the fundamentals of the SBS are described in detail, since it is applied as basic effect to achieve slow- and fast-light in the systems described in this book. The properties of the SBS are explained, including important characteristics such as threshold, Brillouin frequency shift, gain/loss, its bandwidth and limits. In the second section the principle work and setup of a basic SBS-based slow- and fast-light system is explained. Therefore, first investigations and measurements on such a system are presented followed by a summarization of its properties.

The natural limits are caused by the aforementioned disadvantages introduced by the SBS. One aim of this work is to circumvent these limits and to enhance the slow- and fast-light performance. Thus, different opportunities for the enhancement of the Brillouin bandwidth and the time delay as well as the reduction of the pulse distortions of optical single pulses are proposed and discussed in Chapters 4 to 6. In principle, the mitigation of the limitations relies on a shape tailoring of the Brillouin spectrum. A simple modulation of the pump wave can be used for the Brillouin bandwidth enhancement to more than 25 GHz as discussed in Chapter 4. The time delay enhancement described in Chapter 5 can be achieved by a superposition of several Brillouin gain and loss spectra onto an optimized overall spectrum. This leads to a drastic improvement of the relative time delay to four pulse widths. However, the delay is accompanied by a large pulse broadening which restricts the delay measured in units of the output pulse lengths. This is discussed in Chapter 6 and here, it will be shown that the temporal broadening of the pulses can be reduced drastically by broadening the Brillouin spectrum and flattening the group velocity dispersion (GVD). For all these purposes, these three chapters report theoretical treatments and simulations as well as experimental verifications of different Brillouin spectrum schemes. The simulations have been carried out on the basis of a student project work done by A. Wiatrek with the computer algebra system Maple version 11 [19].

During the investigations it was shown that the slow-light effect can be achieved much easier and has a much higher practical importance than the fast-light effect. Therefore, the book focuses on the slow-light effect and only one fast-light scheme is described in Chapter 7. It is also based on a superposition of a Brillouin gain and loss and an advancement of approximately one pulse width could be achieved. This has been examined within an experiment of K.-U. Lauterbach [20].

In Chapter 8 the theoretical and physical limitations of SBS-based slow- and fast-light systems in general are described and the presented schemes are discussed. As it will be shown, the maximum time delay is primarily limited by the saturation and the maximum achievable gain which in turn depends on the signal input power. Therefore, a maximum storage capacity of approximately 4.5 bit can be achieved by the investigated systems for an input signal with a power of -60 dBm and a pulse width of 30 ns.

Based on the results of this work, it is concluded how far these systems are applicable in optical communication and information systems. Therefore, a general summary of the results is provided in Chapter 9. Although the time delay in single stage systems can be multiples of the initial pulse width, large memories are only possible with cascaded zero-broadening systems. Hence, high optical memories are not achievable without suffering from large distortions or a drastic increase in the system complexity. However, synchronizing, equalizing and signal processing as well as phased-array applications can be easily realized and benefit a lot from this effect.

In the last chapter of this book further improvement opportunities of the slow-light performance and future investigations are proposed. This includes the performance enhancement by using other fiber types with an improved Brillouin efficiency for instance, the adaption of the systems on real data streams and the engineering realization of slow-light in existing networks.

2 Fundamentals

This chapter gives an introduction to the fundamentals of slow- and fast-light systems. At first, definitions of different velocities of light are given. A brief description of the KKR as basis to understand the relation between a material resonance and the change of the refractive index follows in the second section. The most important metrics to evaluate the grade and the efficiency of the slow- and fast-light systems are defined in the third section of this chapter. While a verification of the different slow- and fast-light methods is given in Section 2.4, a classification of the main possible applications is provided in Section 2.5.

2.1 Velocities of Light

Before any discussion on the alteration of the speed of light can begin, it is necessary to define different light velocities which are characteristic for this purpose. The importance of that distinction is obvious, particularly in view of the dispute about superluminal velocities and the theory of relativity. Although there are at least eight different definitions of light velocities [21–23], it is sufficient to introduce only four here: the photon velocity, the phase velocity, the group velocity and the signal or information velocity.

2.1.1 Photon Velocity

Every electro-magnetic radiation is quantized in elementary particles, known as photons. These basic 'units' of light move always with same velocity whether they are inside a medium or inside a vacuum. This velocity is determined by the speed of light in a vacuum $c \approx 3 \times 10^8$ m/s [24, p. 29]. Inside a material the photons can be absorbed and re-emitted by its atoms. If this process is linear the frequency of the photons remains constant. Only a phase shift between the absorbed and re-emitted photon occurs. However, in the free space between the atoms, they move with the speed of light in a vacuum.

2.1.2 Phase Velocity

The phase velocity is the velocity at which each point of a constant phase of a wave (e.g. the maximum of a wave) will travel [24]. Assume the complex electric field of a monochromatic wave with a constant angular frequency $\omega = 2\pi f$ as a function of the distance z and the time t is given as:

$$E(z,t) = \frac{1}{2}\left(E_0 e^{j(k(\omega)z - \omega t)} + c.c.\right), \tag{2.1}$$

where E_0 denotes the amplitude of the wave, $k(\omega) = k_0 \times n(\omega) = \omega/c \times n(\omega)$ the frequency-dependent wave number and c.c. is the conjugate-complex part of the wave equation. The phase Φ of this wave is:

$$\Phi(z,t) = k(\omega)z - \omega t. \tag{2.2}$$

Since the phase remains constant while the wave propagates a certain time dt, the phase alteration $d\Phi/dt$ is zero:

$$\frac{d\Phi}{dt} = k(\omega)\frac{dz}{dt} - \omega = 0. \tag{2.3}$$

Hence, the phase velocity is:

$$v_{ph} = \frac{dz}{dt} = \frac{\omega}{k(\omega)} = \frac{c}{n(\omega)}. \tag{2.4}$$

According to Equation (2.4), the phase velocity in a medium is a function of the frequency-dependent refractive index $n(\omega)$ which is referred to as the dispersion relation. For example, in pure silica at frequencies or wavelengths in the range of optical communications $v_{ph} = 2/3\,c$ whereas in a vacuum the phase velocity equals the speed of light c.

If the wave consists of a mixture of different frequencies (a pulse for instance) the different frequency components move with different phase velocities through the dispersive medium. This fact has a large impact on the propagation of the pulse as a whole, because it will not retain its shape. In most materials the refractive index decreases with an increasing wavelength (decreasing frequency). The result is an increasing phase velocity. This case is known as normal dispersion. The opposite behavior, an increasing refractive index with increasing wavelength, is the anomalous dispersion. In most cases, this effect appears in wavelength ranges which are close to resonance wavelengths of the material.

2.1.3 Group Velocity

The velocity with which the pulse, in particular the maximum of the pulse, propagates is called group velocity. As previously mentioned, an optical pulse consists of a group of waves with different frequencies. Hence, it is the modulation of a carrier frequency. Consider two waves with the same polarization and amplitude as Equation (2.1), but with slightly different frequencies and wave numbers. Their phases are:

$$\Phi_1 = (k + \Delta k)z - (\omega + \Delta\omega)t, \tag{2.5}$$
$$\Phi_2 = (k - \Delta k)z - (\omega - \Delta\omega)t. \tag{2.6}$$

The sum of the two waves is then:

$$E_s(z,t) = E_1(z,t) + E_2(z,t) = 2E_0 \cos(kz - \omega t)\cos(\Delta k z - \Delta\omega t). \tag{2.7}$$

The factor $\cos(kz - \omega t)$ is the carrier of the pulse with the phase velocity $v_{ph} = \omega/k$. The second cosine factor gives the wave modulation or 'envelope' of the pulse [5, p. 18], which

2.1 Velocities of Light

moves with the group velocity. The pulses for optical communications consist of more than two frequencies. Since they are composed of a group of waves within a small margin of frequencies and wave numbers around ω and k, $\Delta\omega$ and Δk can also be written as $d\omega$ and dk. Analog to the derivation of v_{ph}, the group velocity v_g is defined as:

$$v_g = \frac{dz}{dt} = \frac{d\omega}{dk}. \tag{2.8}$$

In general, the group velocity depends on the group index n_g which is *the refractive index for the group velocity*. By including Equation (2.4) in Equation (2.8), the group velocity of a pulse with the central frequency ω_0 can also be written as:

$$v_g = \frac{c}{n(\omega_0) + \omega_0 \frac{dn(\omega)}{d\omega}} = \frac{c}{n_g}. \tag{2.9}$$

In a vacuum the group velocity equals the phase velocity and c because $n = 1$ for all frequencies. However, in a medium it will be different from v_{ph}. As can be seen from Equation (2.9), v_g does not only depend on the frequency-dependent refractive index but also on the frequency-dependent slope of the refractive index $dn(\omega)/d\omega$. Hence, only in the case of weakly- or even non-dispersive media $v_g \approx v_{ph}$ and the group velocity is only influenced by $n(\omega_0)$. Thus, the velocity of the pulse can be changed by an alteration of its central frequency for instance.

However, under normal circumstances materials, such as an optical fiber, are indeed dispersive and the change of the pulse's group velocity depends on the refractive index slope. If this slope varies with the frequency the group velocity varies with frequency as well. Therefore, the frequency components of the pulse propagate with different velocities which leads to a broadening of the pulse duration. The frequency-dependent alteration of v_g is the *dispersion of the group velocity* or GVD. It is defined as [24, p. 33]:

$$GVD = \frac{d}{d\omega}\left(v_g^{-1}\right) = \frac{1}{c}\left(2\frac{dn(\omega)}{d\omega} + \omega_0 \frac{d^2n(\omega)}{d\omega^2}\right). \tag{2.10}$$

The normalized group velocity as a function of $\omega_0 dn(\omega)/d\omega$ is shown in Figure 2.1a. In a spectral region of normal dispersion where $dn(\omega)/d\omega > 0$, the group velocity decreases. It is less than the phase velocity and can take on very low values. Since the pulse is decelerated, this corresponds to *slow-light*. If the refractive index slope is negative (anomalous dispersion) the group velocity is increased. Therefore, the pulse will be accelerated and is faster than v_{ph}. This is the region where *fast-light* is the valid expression. In the fast-light regime the group velocity can exceed c. If $|\omega_0 dn(\omega)/d\omega| > n(\omega_0)$ and $dn(\omega)/d\omega < 0$ (for example near material resonances) the group velocity can even become negative. In this case the pulse propagates into the opposite direction.

These different group velocities result in different arrival times of the pulse after a certain length L of the medium. To qualify the arrival time, a pulse delay time t_{del} can be defined as

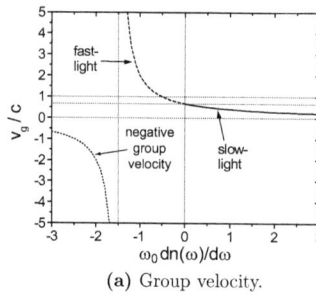 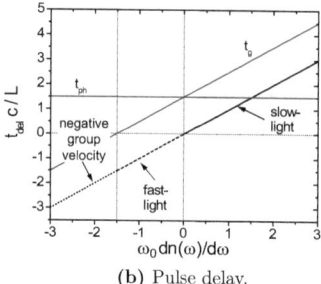

(a) Group velocity. (b) Pulse delay.

Figure 2.1: Group velocity and pulse delay as a function of $\omega_0 dn(\omega)/d\omega$ for $n(\omega_0) = 1.5$. For comparison the solid thin lines show the group and phase delay. In both diagrams the solid bold lines represent slow-light, the dashed bold lines represent fast-light and the dotted bold lines represent negative group velocities.

the difference between the group delay t_g and the phase delay t_{ph}:

$$t_{del} = t_g - t_{ph} = \frac{L}{v_g} - \frac{L}{v_{ph}} = \frac{L}{c} \omega_0 \frac{dn(\omega)}{d\omega}. \qquad (2.11)$$

The pulse delay as a function of $\omega_0 dn(\omega)/d\omega$ in comparison to the phase and group delay is shown in Figure 2.1b. In a non-dispersive medium the pulse propagates exactly with the phase velocity and $t_{del} = 0$. In case of normal dispersion where $dn(\omega)/d\omega > 0$ the group delay is greater than the phase delay. Therefore, the pulse needs more time to propagate along L which results in positive pulse delay values. In the region of anomalous dispersion $(dn(\omega)/d\omega < 0)$ $t_g < t_{ph}$. Hence, the pulse delay becomes negative and the pulse is accelerated. With decreasing values of $\omega_0 dn(\omega)/d\omega$ the group velocity increases rapidly and exceeds also c, as can be seen in Figure 2.1a. If $\omega_0 dn(\omega)/d\omega$ comes into the region of $n(\omega_0)$, $v_g \to \infty$ and t_g becomes zero. The pulse delay equals the value of the phase delay, but with a negative sign. This means that the peak of the pulse arrives at the end of L at the same moment it enters the medium. For lower values of $\omega_0 dn(\omega)/d\omega$ the pulse travels with a negative group velocity and $t_g < 0$. Thus, the negative pulse delay further increases which means that a pulse travels backwards in the medium.

In most cases, superluminal and negative group velocities are possible due to a pulse reshaping inside the medium which results in an advancement of the peak of the pulse. Such a reshaping can be based on an asymmetric absorption or amplification of the pulse energy for instance. Therefore, the pulse's peak is shifted if the trailing edge of the pulse experiences a higher absorption than the front edge or the trailing edge receives less amplification than the front edge [5]. Although a superluminal and negative group velocity is very intriguing and extraordinary, such velocities do not violate Einstein's causality and special theory of relativity. As it will be explained in the next section, the group velocity is not the speed at which an information propagates.

2.1 Velocities of Light

The alterable behavior of the group velocity can be used to engineer systems with large externally controllable dispersions, where $dn(\omega)/d\omega$ has very large positive or negative values. Thus, it was possible to propagate optical pulses extremely fast [25] or extremely slow [26] and even to stop them completely [27, 28]. These results with extreme values of v_g, especially the superluminal velocities, have revived the debates about the velocity of information. In the next section it is more carefully considered what is meant by a *signal* or *information*.

2.1.4 Signal or Information Velocity

In the previous section it has been shown that the group velocity can exceed c. This fact has already been known for more than a century and at the beginning it was believed that v_g is equal to the velocity of information [5]. Therefore at first, there was no doubt about superluminal information which would allow to observe an effect before its cause is happen. However, this was in strict contradiction with Einstein's theory of causality. Therefore, with the formulation of Einstein's special theory of relativity, which says that no signal can move faster than c, the meaning and definition of the velocity of information or signal had to be resolved.

With the work of Sommerfeld and Brillouin [6, 7] a distinction between the group and the information or signal velocity was introduced. They declared that these velocities are identical for nonabsorbing media, but in general, the group velocity is not a signal velocity. In contrast to the other velocities, the meaning of the signal velocity is difficult to define because of the difficulty to describe what an *information* represents. In general, a *signal* is the carrier of the information. The signal involves only new information or information content if an element of *surprise* is transmitted that could not be predicted earlier. According to this, the signal wave should necessarily contain a discontinuity, such as a sharp wavefront that jumps from zero to a finite value for instance [5]. Such a signal can be provided by step function behavior or even the front of a pulse. In agreement with this assertion, Brillouin and Sommerfeld defined a signal as *limited wave motion* which starts abruptly at a certain moment in time and stops after a certain time [7]. Today, this is called a *non-analytical point* [29], as described later in this section.

This means that the peak of a pulse, which is related to the group velocity, does not contain any information because of the absence of a discontinuity. The information is rather located at the front of the pulse. Hence, the signal velocity is related to the pulse's front velocity which is determined by the finite frequency response of the propagation medium [5]. The reduced light speed and the dispersion in a material is based on a rhythmical oscillation of the particles or electrons with the wave. Originally, the electrons are calmly if the thermal dithering is neglected. Hence, the medium can be seen as optically empty or as a vacuum for the front of the pulse. Its light velocity would be c and the refractive index would be $n = 1$. However, the occurrence of an incident wavefront begins to displace the particles of the medium from their initial positions and forces them to oscillate. This motion causes a scattering of radiations from the incident wave. The radiated fields of all particles superimpose coherently to a new field. Due to the inertia of the particles they cannot respond instantaneously to the wavefront.

Therefore, the new field propagates more slowly and its velocity cannot exceed c [2]. Thus, the front velocity is restricted and identical to c. While the group velocity can become faster than the speed of light in a vacuum, the signal or information velocity is always bound to c [29, 30]. Hence, there is no violation of the principle of causality and theory of relativity. The distant leading edge of the pulse contains all information about the entire pulse shape which can be fully reconstructed before the input pulse's peak and trailing edge enters the medium [31].

The front velocity of pulses are defined in different ways. For example, Sommerfeld and Brillouin defined it as *the velocity of the main body of the pulse*, more precisely as *the velocity of the first point which reaches the half of the pulse's maximum intensity* [5, 7, 32]. However, according to the previous statement, that the whole information is already contained in the leading edge of the pulse, this point does not provide any new information and therefore, it cannot be characterized as a real signal. Today, the velocity of the main body of the pulse is rather related to the group velocity.

Moreover, as also obtained by Sommerfeld and Brillouin, a signal does not arrive suddenly. Therefore, there would be a quick but still continuous transition from very weak intensity to that corresponding to the signal [5]. The weak intensity is also called forerunner or precursor. At first a set of precursors occur after the distance z at $t = z/c$, since the medium cannot respond instantaneously to the applied field. The precursors consist of high and low frequency fluctuations with a small amplitude and move with a velocity equal to c. Because of the arising forced oscillation of the electrons their amplitudes increase at $t = z/v_{ph}$. The steady state where the free oscillation of the particles is subsided and the main part of the pulse is formed is reached at $t = z/v_g$ [2]. Thus, the wavefront is related to the precursors and the front or leading edge of a pulse can be defined as the first point where the electromagnetic field is non-zero [32].

Another more specific definition of the wavefront and information velocity has been proposed by Chiao and Steinberg [29]. It says that the information is included in all points of non-analyticity. These points are completely random and cannot be desired by analytical functions. The domain before them is zero and a prediction of the waveform is impossible. After the non-analytical points appeared the behavior of the whole waveform can be determined and becomes analytic immediately. This approach is quite similar to the first-non-zero-point definition. The non-analytic points can also be seen as the wavefront and they propagate with a velocity equal to speed of light in a vacuum, so that the information velocity is c too [32].

Currently, it is generally accepted that the information velocity cannot exceed c. However, within the research community the discussion about its definition and the interpretation of the results of several experiments with superluminal velocities probably has not been finished yet [33]. Although this is a very interesting topic, it is not necessary to explain the signal velocity in more detail here because superluminal velocities are not subject of this work. More information on the signal velocity topic can be found in [5, 29, 32].

2.2 Refractive Index and Kramers-Kronig Relations

As has been explained in the previous section, the change of the velocities of light is bound to oscillations of material particles. This process is represented by the refractive index n which depends on the material itself. The influence of the material is expressed by the relative permittivity ε_r and the relative magnetic permeability μ_r [34]. Then, the refractive index can be written as:

$$n(\omega) = \sqrt{\varepsilon_r(\omega)\,\mu_r(\omega)} \quad \text{with} \quad \varepsilon_r(\omega) = 1 + \chi(\omega), \tag{2.12}$$

where $\chi(\omega)$ denotes the frequency-dependent electric susceptibility. The electric susceptibility is a material property which describes the ability of polarization of the medium inside an electrical field. In dispersive dielectric media $\mu_r(\omega) \approx 1$ and the refractive index, the permittivity and the susceptibility are complex. Therefore, the complex relative permittivity becomes:

$$\hat{\varepsilon}_r(\omega) \approx \hat{n}(\omega)^2 = (n'(\omega) + jn''(\omega))^2 = n'^2(\omega) - n''^2(\omega) + 2jn'(\omega)\,n''(\omega), \tag{2.13}$$

where the real part $\Re\{\hat{\varepsilon}_r\}$ leads directly to the refractive index n and the imaginary part $\Im\{\hat{\varepsilon}_r\}$ leads to an absorption α. For transparent materials which are used for light propagation, $n'' \ll n'$, so that the term $n''^2(\omega)$ can be neglected in Equation (2.13) [34].

As previously described, both the refractive index n and the absorption coefficient α are frequency-dependent which is known as dispersion. Additionally, both quantities are related to each other directly. In 1928 Kramers and Kronig showed that there are integral connections between the real and imaginary parts of the complex linear susceptibility, $\chi'(\omega)$ and $\chi''(\omega)$ [35], which henceforth was called KKR. The KKR is a special case of the Hilbert-transform which in general relates the real and imaginary components of transfer functions of linear causal systems. Hence, if the real or imaginary part of such a function is given the complementary part can be determined. This means simultaneously that a material dispersion leads to an absorption and vice versa [34, 36].

The most common expression of the KKR is [18, 23, 36–39]:

$$\chi'(\omega) = \frac{2}{\pi} \int_0^\infty \frac{\omega' \chi''(\omega')}{\omega'^2 - \omega^2} d\omega', \tag{2.14}$$

$$\chi''(\omega) = -\frac{2}{\pi} \int_0^\infty \frac{\omega \chi'(\omega')}{\omega'^2 - \omega^2} d\omega'. \tag{2.15}$$

In practice it is more useful to work with modified versions of Equation (2.14) and Equation (2.15) which directly relate the refractive index $n(\omega)$ and the absorption coefficient $\alpha(\omega)$ [23, 34, 36, 39]:

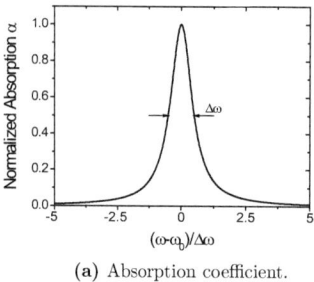
(a) Absorption coefficient.

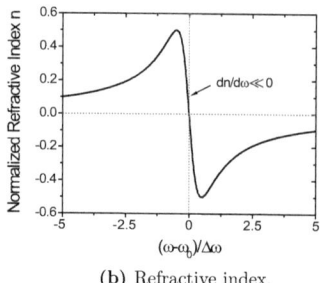
(b) Refractive index.

Figure 2.2: Absorption coefficient and refractive index as a function of the frequency detuning normalized to the absorption FWHM bandwidth $\Delta\omega$. Both, α and n are related through the KKR.

$$n(\omega) = 1 + \frac{c}{\pi} \int_0^\infty \frac{\alpha(\omega')}{\omega'^2 - \omega^2} d\omega', \tag{2.16}$$

$$\alpha(\omega) = -\frac{4\omega^2}{\pi c} \int_0^\infty \frac{n(\omega') - 1}{\omega'^2 - \omega^2} d\omega'. \tag{2.17}$$

The relation between the refractive index n and a Lorentzian-shaped absorption[1] with a FWHM bandwidth of $\Delta\omega$ as a function of the normalized frequency detuning $(\omega - \omega_0)/\Delta\omega$ is shown in Figure 2.2. Within the FWHM bandwidth the absorption spectrum leads to a strong anomalous dispersion where the refractive index has a negative slope. Hence, according to Equation (2.9) the group index will be decreased and the group velocity will be increased. In contrast to this, a peak or gain spectrum would lead to a strong normal dispersion where the group index is increased and the group velocity is decreased. Thus, the KKR show that a strong material resonance, which results in amplification or absorption processes, produces a large dispersion which is necessary for slow- or fast-light. Such a feature is the basis of many methods for the group velocity alteration.

2.3 Figures of Merit and Metrics

To evaluate the grade of delay and the efficiency of the slow- and fast-light systems there are some figures of merit and metrics necessary. Some of them are described and defined in the following paragraphs. In this book the primarily used parameters are the absolute time delay, the fractional time delay, the effective time delay and the broadening factor.

---

[1] The absorption coefficient must contain negative values for processing the KKR to achieve the correct dispersion behavior.

2.3 Figures of Merit and Metrics

2.3.1 Group Index, Group Velocity, Group Delay, Time Delay, Slowdown Factor

As already introduced in Subsection 2.1.3, the group index and the group velocity are the basis for the time delay. Both are equivalent measures which are connected by c to $v_g = c/n_g$ (as shown by Equation (2.9)). In practice, they are usually determined by the propagation time of the pulse through the medium with a certain length, mostly expressed as group delay $t_g = L/v_g = Ln_g/c$. For the propagation through a waveguide without any influence of other slow- or fast-light effects t_g is equal to the absolute time delay ΔT. However, by slowing down or accelerating the pulse with a slow- or fast-light method, as described in Section 2.4, the group delay will change. Then, the time delay is rather defined as the difference between the 'new' group delay with slow- or fast-light effect and the 'normal' group delay without these effects. In practice this difference can easily be determined by measuring the time difference between the peak of the delayed or advanced output pulse and the peak of a reference output pulse without applying a group velocity alteration.

The grade of the group velocity change is commonly expressed by the slowdown factor S which relates a reference velocity, e.g. the phase velocity or c, to v_g [40–42]:

$$S = \frac{c}{v_g} \equiv n_g \quad \text{or} \quad S = \frac{v_{ph}}{v_g} = \frac{n_g}{n(\omega)}. \tag{2.18}$$

If S is related to c the slowdown factor directly gives the group index. Furthermore, S shows directly the deviation of the propagation velocity in the material to that in a vacuum. Thus, it can be evaluated immediately if v_g is superluminal or below c. Contrary to that, the latter expression of Equation (2.18) focuses on the slow- or fast-light effect itself by normalizing the group velocity to the velocity obtained in the unstructured material [42]. Therefore, S equals the ratio of n_g over the material refractive index n.

However, these simple parameters show the performance of a slow- or fast-light system insufficiently if they are mentioned alone. For example, the absolute time delay measured at the peaks of the pulses does not declare anything about possible distortions or temporal broadenings of the delayed or advanced output pulse. Furthermore, the value of the time delay is only meaningful if the appropriate temporal FHWM width or bit rate is known. A ΔT of 10 ns is large for pulses with widths of 1 ps but insignificant for 1 µs pulses for instance. Hence, an indication of additional measures is necessary.

2.3.2 Fractional and Effective Time Delay

The fractional time delay ΔT_{frac} is defined as the ratio between the absolute time delay ΔT and the input FWHM width τ_{in} of a single pulse, whereas the effective time delay ΔT_{eff} refers to the output pulse width τ_{out}.

$$\Delta T_{frac} = \frac{\Delta T}{\tau_{in}}, \qquad (2.19)$$

$$\Delta T_{eff} = \frac{\Delta T}{\tau_{out}}. \qquad (2.20)$$

These two parameters are more adequate when evaluating the time delay because they are related to the pulse width. Therefore, for the examples from Subsection 2.3.1 fractional delays of 10^4 and 10^{-2} are achieved. Although the same absolute time delays were assumed the fractional time delays show a significant difference. Furthermore, ΔT_{frac} can be used to predict the theoretic storage or delay capacity in [bit].

If the pulse width is maintained in the system the effective time delay is equal to the fractional time delay. However, if the pulse is distorted, i.e. compressed or broadened, ΔT_{eff} deviates from ΔT_{frac} and shows the 'real' time delay of the pulse. Hence, it takes into account every pulse distortion which influences the temporal FWHM pulse width. In most cases, the pulse width is broadened due to the large dispersive behavior of the slow- and fast-light effect. Thus, the effective time delay will be reduced and is less than ΔT_{frac}.

2.3.3 Broadening Factor

The broadening factor B describes the grade of compression or broadening of the pulse after slowing down or advancement. It is defined as the ratio between the temporal output and input FWHM pulse widths:

$$B = \frac{\tau_{out}}{\tau_{in}} = \frac{\Delta T_{frac}}{\Delta T_{eff}}. \qquad (2.21)$$

Therefore, it is directly connected with ΔT_{frac} and ΔT_{eff} and leads to a limitation of the time delay. By using B, the maximum allowed broadening can be considered for a certain system. Mostly, B should be ≥ 1 and < 2, with $B = 1$ as the optimum. If $B > 2$ a loss of information inside a bit stream takes place due to intersymbol interferences [43].

2.3.4 Delay-Bandwidth Product, Delay-Bit Rate Product

The delay-bandwidth product or delay-bit-rate product (DBP) is a valuable parameter which combines the time delay of the pulse or bit stream with its bandwidth (or optical bandwidth of the data channel) or bit rate. It is a constant measure for the system and can be expressed by [44]:

$$DBP = \Delta T \, \Delta f, \qquad (2.22)$$

where Δf denotes the FWHM bandwidth or bit rate and ΔT is the time delay for each pulse in the sequence. In most slow- and fast-light systems large absolute time delays are accompanied by low bit rates which leads to small fractional delays. The opposite is that a high bit rate with extremely small bit times experiences only small delays, but the fractional delay can be higher. However, if only single pulses are used, as in this work, then the temporal width of the pulse is $\tau_{in} \approx 1/\Delta f$ for ideal rectangular pulses or $\tau_{in} \approx 0.441/\Delta f$ for Gaussian-shaped pulses. Therefore, $DBP = \Delta T_{frac}$ and $DBP = 0.441\,\Delta T_{frac}$, respectively.

2.3.5 Q-Delay Product

One opportunity to quantify the data distortion after the transmission is the Q factor which is directly related to the signal-to-noise-ratio, the bit error ratio (BER[2]) and the eye diagram opening. It decreases with an increasing distortion. The Q-delay product combines the amount of distortion with the delay and thus, provides an insight into the optimal design of the slow- and fast-light system [44]. For example, if the Q-delay factor is considered to be a function of the ratio between the signal bandwidth (or bit rate) and the resonance or system bandwidth $\Delta f / \Delta \omega$ these two bandwidths can be adjusted to find an optimal operating point where the delay is as high and the distortion is as low as possible [4, p. 329]. On the one hand, for $\Delta f \ll \Delta \omega$ there is almost no distortion but also a negligible delay. On the other hand, if $\Delta f \gg \Delta \omega$ this results in a high delay but also a high distortion. Thus, the Q-factor product can be used to find a tradeoff between the delay and the distortion.

2.3.6 Other Metrics

There might be more figures of merit which should be carefully and individually examined for the different slow- and fast-light systems. Each application imposes its own metrics that must be met. Some further typical delay metrics are listed below [4, p. 325].

BER and Eye Opening The BER is very important if digital information are transmitted and can be verified from the opening of an eye diagram. For an error free transmission the BER usually should not exceed a value of 10^{-6}. Some systems actually need BER values less than 10^{-9} or 10^{-12}.

Delay Range The delay range is the tuning range of the time delay between its minimum and maximum value.

Delay Resolution The delay resolution is the minimum incremental delay tuning step.

Delay Accuracy The delay accuracy denotes the precision percentage of the actual delay value to that of a desired one.

Reconfiguration Time The reconfiguration time defines the amount of time which is necessary to switch the time delay from one to another steady state value.

---

[2]The BER indicates the ratio of the received numbers of bit errors over the total number of bits within a certain time interval. It can be estimated from an eye diagram opening for instance.

Absorption and Loss over Delay The absorption is the total amount of loss induced to the signal during the delay or advancement process, whereas the loss over delay determines the amount of loss incurred per unit time delay.

2.4 Slow- and Fast-Light Methods

This section gives an overview of the most common slow- and fast-light methods. At first, two possible classifications are introduced. In the following paragraphs the several slow- and fast-light techniques are described in more detail.

During the past few years various methods, mechanisms and materials to achieve a delay or acceleration of optical pulses have been proposed. As described in Section 1.1, they can be separated into several categories. There are three basic techniques:

1. methods which rely on the physical length of a waveguide or resonator,

2. methods which are based on a change of the group velocity of the signal pulse and

3. methods on the basis of the time-frequency-coherence of signals.

Examples of these techniques are itemized in Figure 2.3. In general, the first and second mechanism belong to the propagation of the pulses with group velocity v_g in media. In the first method, the alteration of the arrival time of the pulses is achieved by changing the physical length of the propagation path. For this, either the length of a single waveguide can be directly altered or the cycles of the pulse inside a closed waveguide-loop [40, 45] or resonator [46–49] are changed. The disadvantage of this method is that in most cases the time delay is only a multiple of the bit time and cannot be varied continuously.

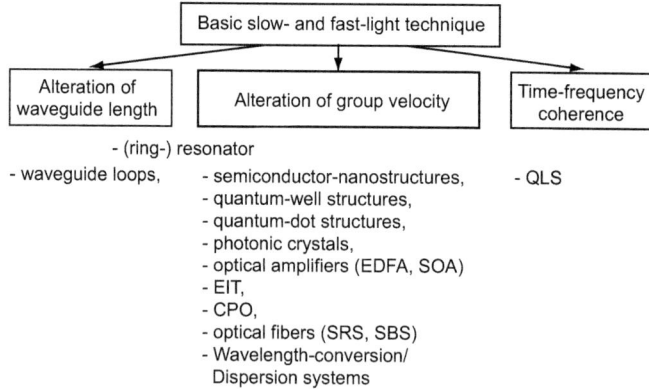

Figure 2.3: Classification of the slow- and fast-light methods according to the basic slow- and fast-light technique.

2.4 Slow- and Fast-Light Methods

While the group velocity of the pulse remains the same in the first method, in the second technique v_g is altered by itself whereas the length of the material remains the same to achieve the time delay. This can be realized using methods or materials which provide a sharp resonance or dispersion, such as ring-resonators [46–49], semiconductor-nanostructures [50], quantum-well [51, 52] and -dot structures [52, 53], photonic crystals (PC) [41, 54–56], optical amplifiers (e.g. erbium-doped fiber amplifiers (EDFA) [57, 58] and semiconductor optical amplifiers (SOA) [59–64]) as well as electromagnetically induced transparency (EIT) [26, 65, 66], coherent population oscillations (CPO) [57, 67–70], nonlinear effects in optical fibers (e.g. stimulated Raman scattering (SRS) [71], SRS-assisted optical parametric amplification [72] or SBS [13–15, 17]) and systems which connects different dispersion features with a wavelength-conversion [73–81].

The third class includes a novel technique called quasi-light storage (QLS) [82]. This method exploits the coherence between the time and frequency representation of a signal. The big advantage of QLS is that light pulses or pulse bursts can be stored for a certain time with almost no distortion.

Another classification can be used according to the nature of the propagation time alteration of the pulse which is shown in Figure 2.4. On the one hand, it is possible to influence the velocity, the propagation time or the propagation path of the light pulses. These methods belong to slow- and fast-light. Most of them are able to change the group velocity in very slight steps and hence, provide a fine tuning of the delay or acceleration. The others only achieve fixed delay values which correspond to multiples of a single transmission time through the system. Therefore, they are only useful for a coarse tuning. On the other hand, the light pulses can be

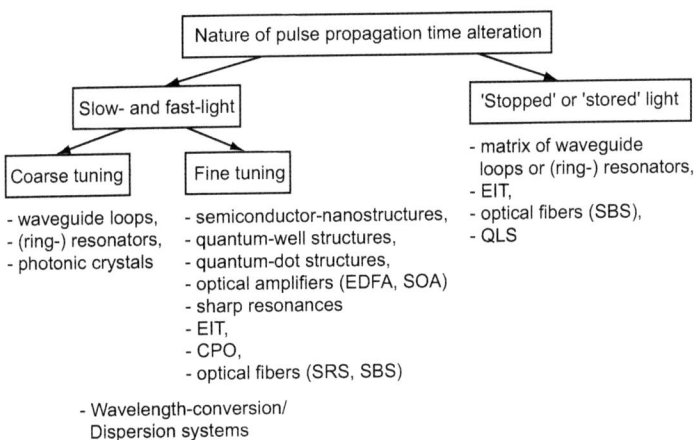

Figure 2.4: Classification of the slow- and fast-light methods according to the nature of the pulse propagation time alteration.

completely stopped or stored. In contrast to a change of the pulse propagation time, the signal remains inside the memory until it will be released by a read signal. For this purpose also some of the slow- and fast-light methods are usable.

2.4.1 Waveguide-Loops

The easiest way to realize an optical pulse delay is to send the signal into a delay line with a fixed length L, e.g. an optical waveguide or fiber [40]. Inside the fiber segment the pulse propagates with the group velocity which depends on the group index of the transmitting medium. The time delay is then caused by the fiber length and is equal to the group delay $t_g = L/v_g$. If higher delays are needed the delay line can also be passed through several times. This leads to multiple delays of $i \times t_g$ with i as the number of pass through the delay line. The pulse propagates as long in the loop as it will be switched onto another path. Hence, the signal can be stored a certain time or arbitrarily long if the waveguide attenuation is neglected.

A further development of this method is to use a network of waveguides with different lengths which are connected by switches [45]. By using different combinations of the waveguide or fiber loops, different delay times can be implemented. For example, such a network was used to achieve a delay up to 2.56 ns [83].

Although this method is very simple it has some serious disadvantages. The achievable time delay or storage time depends directly on the constructional defined length of the waveguide. If the signal is coupled into the loop once it can be read out only after a full circulation. Hence, the time delay cannot be varied. However, for most applications of an optical delay line (as described in Section 2.5) a variable time delay is indispensable. Another problem is the attenuation of the waveguides which strongly reduces the signal power if L or i take on very large values. Additionally, such high values also induce a high signal distortion due to the waveguide dispersion. At last, a large network of waveguide-loops needs a lot of space if it is built up with fiber coils. For example, if single-wavelength optical fiber delay lines were used to buffer 1000 channels, each with a data rate of 40 Gbit/s and with a storage capacity of 10 Gbit, fiber loops with a total length of approximately 150 times the distance from the earth to the moon would be necessary [40].

2.4.2 Electromagnetically Induced Transparency and Coherent Population Oscillations

It is already well known for a long time that the group velocity is changed near a material resonance. Thus, the first experiments on slow- and fast-light were done at strong absorption resonances which can be found for example in dilute gases [84]. On the one hand, such a gas has typically a very small refractive index of the order of $n - 1 \approx 10^{-3}$. On the other hand, at the spectral boundaries of the absorption it varies and presents a normal dispersion, as can be seen in Figure 2.2b. Hence, the group index can be very large, in the order of 10^6, which leads

2.4 Slow- and Fast-Light Methods

to $v_g \approx 300\,\text{m/s}$ [84]. However, due to the absorption resonance the attenuation is very high and the light is totally absorbed after a short distance.

By using the coherent quantum interference effect of EIT, this drawback can be canceled while the slow-light effect is preserved [85]. Via EIT a narrow transparency window within the absorption profile can be created, as shown in Figure 2.5a. A pulse within this window experiences little or no absorption and is associated with a change of its group velocity. According to the KKR the transparency window leads to a rapid variation of the refractive index and therefore, to a large normal dispersion, as can be seen in Figure 2.5b. Hence, the magnitude of the group velocity change and the time delay depends on the efficiency of the transparency and on the slope of the dispersion, respectively.

The effect of EIT was first discovered by Harris et al. in a gas of atoms that have three energy levels [86, 87]. A weak probe field was tuned near one of the transition wavelengths ($\lambda = 283\,\text{nm}$) of the absorption resonance. A second much stronger coupling (also called control or pump) field was tuned to the other transition wavelength ($\lambda = 405.9\,\text{nm}$), so that it induces a coherence between the ground state and the excited level. This leads to a destructive quantum interference between both transitions and to a narrowband cancellation of the absorption [4, 44].

EIT was used in 1999 by Hau et al. to slow down light pulses to a velocity of $17\,\text{m/s}$ which is equal to the speed of an average bicyclist [26]. In that experiment they applied a strong coupling field to an ultra cold atomic cloud in the form of a Bose-Einstein condensate[3] to create the narrow transparency window within the absorption resonance. This result was the main initiation for the reinforced research on the topic of slow- and fast-light. Later on, the group of Hau succeeded to stop the light pulse for $1\,\text{ms}$ [27]. Recently, they even showed in a noticeable experiment that it is possible to stop the light pulse fully in one Bose-Einstein condensate and revive it after a short time in another condensate which was spatial separated by $160\,\mu\text{m}$ [88]. This was possible due to the quantum mechanical indistinguishability of both Bose-Einstein condensates.

Also hot atomic gases can be used for achieving slow-light based on EIT. For example, a group velocity as low as $90\,\text{m/s}$ [65] and a temporary light storage [89] were shown in a coherently driven hot gas of rubidium atoms with a temperature of around $360\,\text{K}$. Furthermore, EIT-based slow- and stopped-light have been demonstrated in solids, such as crystals [66, 90] and semiconductors [44].

Although the results of the EIT are noticeable, EIT suffers from significant disadvantages. First, the pulse wavelength has to be adjusted exactly to the wavelength of the material resonance. In most cases, these wavelengths are outside of those which are used in optical communications nowadays. Furthermore, the resonances and hence, the transparency windows are extremely narrowband. Therefore, it is not possible to cover todays common bandwidths of tens of Gbit/s. Another major problem for a practical application of EIT-based slow-light

---

[3] A Bose-Einstein condensate is a state of matter of a dilute gas of weakly interacting indistinguishable particles cooled down to temperatures very near to absolute zero ($0\,\text{K}$). Under such conditions the particles have the same quantum mechanical state and their occurrence probability is absolutely equal all over the condensate. Thus, the state can be described by only one wave function.

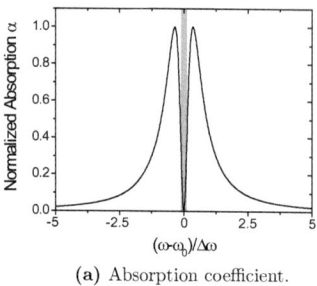 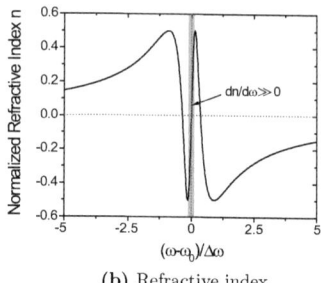

(a) Absorption coefficient. (b) Refractive index.

Figure 2.5: Absorption coefficient and refractive index as a function of the frequency detuning for an EIT scheme. The shadowed area shows the narrow transparent spectral range with normal dispersion.

systems is their difficult handling because the exotic materials need extreme temperatures to avoid any disturbance of the quantum interference by other effects and to obtain a large group index. For these reasons, alternative methods have been quested which work in solids at room temperatures and at wavelengths or frequencies used in optical communications.

A method which has a similar behavior like EIT but works in solids at room temperature is the concept of CPO. It leads also to a narrow spectral window inside an absorption profile and therefore, causes a change of the group index. However, while EIT involves a quantum mechanical interference effect between the electronic state wave functions, the coherence for CPO is assured by the interference of two external laser beams [70]. This quantum effect occurs in materials with two energy levels and which show a saturable absorption. Near an allowed broadband transition a weak probe wave and a pump wave with slightly different frequencies are applied to the medium. If both fields are nearly resonant and the pump is strong enough their interaction causes the atomic population to oscillate between the ground and excited state at the beat frequency between the probe and the pump. The result is that the light is efficiently scattered from the pump into the probe and hence, the absorption is decreased [4, 5, 44].

The great advantage of CPO is that it can be realized in a variety of materials at room temperatures and at wavelengths used in optical communications. Ultra slow- and fast-light and even negative group velocities have been shown in crystals, such as ruby [67] and alexandrite [68], as well as in an erbium-doped optical fiber at pulse wavelengths of 1550 nm [57, 69]. Furthermore, CPO-based slow-light was achieved in semiconductor structures [70].

Although CPO seems to be a very promising slow- and fast-light technique it still suffers from a narrow bandwidth which is set by the inverse of the population recovery time. For crystal and erbium structures it is restricted to a few kHz which allows to only use pulses much longer than 1 ms [84]. With semiconductor structures it is possible to achieve much faster population recoveries and therefore, higher bandwidths on the order of a few GHz [51, 91]. However, promising slow- and fast-light techniques should be capable of delivering tens of GHz.

2.4.3 Engineered Materials and Structures

Arbitrary resonances can be efficiently generated in engineered materials or structures, such as periodical structures, semiconductors and crystals. They offer the opportunity to design fortunate resonance and dispersion properties which are easily controllable by the bias voltage or injection current as well as the optical pump waves themselves. The vast potential advantages include room temperature operation, low power consumption, compactness and opportunity for monolithic on-chip integration [50].

Large positive waveguide dispersions are achievable by periodical structures or reflections on a grating, e.g. fiber Bragg grating (FBG) [92, 93]. In practice however, mostly ring-resonators, known as coupled resonator optical waveguides (CROW), are used because they can offer good performance in the range up to 1000 Gbit/s [46]. Ring-resonators are closed loop optical waveguides with in- and outputs. Light of a proper wavelength circulates within the loop and build up a constructive interference before coupled out. The interference leads to a resonance which can be used for slow-light. CROW on a glass substrate enables high time delays with high bandwidths at low signal attenuations < 1 dB/bit on physically small dimensions. For example, a delay of 100 ps Gaussian-shaped pulses about 8 bit has been demonstrated on an area of 7 mm^2 [49]. Ring resonators on the basis of silicon-on-insulator (SOI) structures offer also a high time delay with ultrahigh bandwidth or even light storage [47], but on a smaller size. With such a SOI structure inside a nanoscale waveguide signal pulses as short as 3 ps have been delayed about 4 ps using SRS [94]. An error-free 2.5 bit delay of data rates up to 5 Gbit/s has been shown in a device with a footprint below 0.09 mm^2 [48].

One significant problem of the SOI structures is their large signal attenuation at the resonance wavelengths. The device presented in [48] has an insertion loss of 22 dB. Another drawback is that the time delay, like that for waveguide-loops, can only be a multiple of a certain value which depends on the construction of the ring-resonator. Thus, it can only be tuned in a coarse range. Although a fine tuning via heating of the substrate have been proposed, this is very complicated and too slow [95].

Semiconductor-nanostructures also suffer from high absorption rates which prevent a scaling of the time delay with the device length. However, in contrast to CROW, semiconductor-nanostructures enable a fine tuning of the pulse delay. The refractive index of the semiconductor is changed by an alteration of absorption or gain spectra, mostly induced via EIT or CPO [4, 50]. The available bandwidth is determined by the population lifetime equal to the radiative recombination lifetime which can be in the range of hundreds of picoseconds up to 1 ns [4, p. 14]. Hence, data rates up to tens of Gbit/s can be processed.

For slow- and fast-light quantum-dot or quantum-well structures are usable [52]. These structures are tiny artificial semiconductor crystallites that confine the charge carriers in one direction or in three dimensions [50]. This results in discrete and not continuous energy states where the recombination of the charge carriers define the transmission spectrum, as aforementioned. The quantum mechanical states and therefore, the optical properties of the quantum-dots and -wells can be tailored by designing their form, size and number. For example a group velocity reduction to 9600 m/s has been achieved within a transparency window which exhibits a FWHM bandwidth of 2 GHz in a quantum-well structure [51]. Both slowdown and advancement of 170 fs pulses with fractional delays from −0.2 up to 0.4 have been shown by optical and electrical tuning of quantum-dot SOA at room temperature [53].

In recent years, further engineered artificial structures, which have become very attractive for generating slow- and fast-light, are PC devices. They work also with an arising dispersion and can offer wide-bandwidth and dispersion-free propagation at room temperatures. PC are multidimensional arbitrary periodic structures with periods of the order of the wavelength range [56]. Nowadays, PC devices mostly consist of thin films with two-dimensional arrays of air holes surrounded by air claddings. By now, their manufacture is simple and they benefit from their almost lossless light propagation.

The dispersion characteristics of PC is based on the existence of a photonic band gap, a line defect of missing air holes in the PC waveguide (also called stop band), which causes frequency bands of inhibited optical modes. If the light propagates through the defect it will be totally reflected in vertical direction and Bragg reflected in lateral direction. This leads to a strong dispersion and hence, to a group velocity alteration near the photonic band edge [4, Chapter 4], [56]. Although PC devices provide a high bandwidth capability, a tuning of the time delay remains difficult because the band gap is set by the manufacturing process which makes a change afterwards very complicated. In PC waveguides group velocities as small as $c/30$ to $c/1000$ have been observed for picosecond pulses [41, 54, 55].

2.4.4 Optical Amplifiers

Besides absorption resonances, amplification processes are also applicable to induce a material dispersion. Hence, optical amplifiers can be used for slow- and fast-light. Optical amplifiers benefit from the opportunity of the integration on a small physical size and the capability of high bandwidth while the maximum achievable time delay is small compared to other mechanisms. As every amplifier they also suffer from an increased amplified spontaneous emission (ASE) with the amplification process.

A fractional advancement of approximately 0.15 of amplitude-modulated 1550 nm signals has been shown in an EDFA [58]. SOA can used for a time delay of light pulses as well. In contrast to absorbers, where the maximum delay is related to the decay of pump power along the device, in a SOA the delay time increases with the device length and is limited by the corresponding growth of the gain as well as of the ASE. This restricts the number of delayed bits to a value between one and two [59]. By concatenating multiple amplifying and absorbing sections large phase shifts between 110° and 150° can be achieved with small distortions [61, 96]. A radio-frequency phase shift $> 360\,°$ over several tens of GHz has been also demonstrated by cascading optical filtering amplifiers [64]. Also fast-light can be obtained using multiple cascaded quantum-well SOA [60]. In [62] a time delay of 0.4 ns is reported by applying four-wave mixing (FWM) in a SOA. That SOA enables ultrahigh bandwidth slow- and fast-light has been shown in [63] where amongst others a 190 fs pulse has been tunable advanced by 0.71 ps.

2.4.5 Optical Fibers

Also optical fibers can be used efficiently for the group velocity alteration. The great advantage of fibers is that they can act as slow- or fast-light medium directly. Thus, they are insertable into (existing) optical information and communication systems seamlessly. Furthermore, there are many inexpensive and reliable fiber types with different properties available that makes them very flexible to apply. Another important feature is that the slow- and fast-light effect occurs at room temperatures and within a large wavelength range, especially in those which are used in optical communications.

For example, a continuous control of delays greater than 800 ps near 1550 nm for a wide range of signal pulse durations (ps to 10 ns) has been presented by a combination of FWM and fiber dispersion [73]. Such a conversion-dispersion-technique will be explained in more detail later in Subsection 2.4.6.

To achieve slow- and fast-light in an optical fiber stimulated scattering processes, such as SRS and SBS, are of special interest. Thereby, the change of the group velocity is based on artificial created resonances in the medium. Since SRS and SBS are induced by a pump wave, they can occur over the entire transparency range of the fibers. Additionally, a design of the spectral resonance shape by the pump provides an easy tailoring of the dispersive properties of the optical material.

In stimulated scattering processes the photons of a light wave interact with the medium. They deliver motion energy to or absorb it from the medium. This is connected with a change of both, the frequency and the direction of the incident wave [24, p. 242]. Sufficiently strong light waves induce a material excitation or resonance which is coupled to the light fields if their frequency difference is equal to the frequency of the excitation. This gives a rise to the nonlinear coupling between the waves and the energy can flow from one to another [4, p. 38]. Hence, a probe wave can be amplified or absorbed. Additionally, according to the KKR, this leads to a change of the refractive index with the frequency and therefore, to slow- or fast-light.

In SRS, the scattering arises from the interaction between the pump wave and exciting vibrational motions or oscillations of the individual particles, also known as phonons. In a silica fiber the SRS gain (Stokes[4] band) or loss (anti-Stokes band) occur about 13.2 THz below

---

[4]The wave which caused the scattering is called pump wave, scattered waves with a lower frequency are Stokes-waves, whereas scattered waves with a higher frequency are called anti-Stokes-waves [24, p. 242].

or above the pump frequency. The linewidth of the resonances is very high and can be in the range from a few THz up to 40 THz. With SRS ultrahigh bandwidth slow-light has been shown where 430 fs pulses were delayed by up to 85 % of the pulse width [71]. In [72] a Raman-assisted fiber optical parametric amplification in a fiber with a length of 2 km was used to achieve negative delays as well as large positive delays and delay tuning ranges, in the order of 160 ps for a 70 ps wide pulse.

As opposed to SRS, the scattering of SBS is based on acoustic oscillations or acoustic waves. During the last years, SBS has become a very promising fiber-based technique to realize slow- and fast-light [13, 14, 17]. This is due to its crucial advantages over the other methods. With SBS it is possible to tune the group velocity continuously in an extremely wide range. In short SSMF of 2 m group velocities of less than 71 000 km/s up to superluminal and even negative group velocities were achieved [15]. It was also shown that SBS-based slow- and fast-light can be used to realize potential applications in the field of optical communications, such as all-optical synchronization and multiplexing [4, Chapter 16]. Moreover, such systems might be useful in future heterogeneous multiformat, multibit-rate, and multichannel fiber-optic communication or information systems and networks [97]. Since, SBS is the key effect used to produce slow- and fast-light in this work, it is described in detail in Chapter 3 and further discussed in the following chapters.

Recently, it has been reported that SBS can also be used to stop sequences of light pulses completely and to store them for a certain time [28]. To implement this, the information contained in the optical pulses were converted into long-lived acoustic excitations in an optical fiber. A write pulse, counter-propagating at a frequency downshifted from the data pulse frequency, depletes the data pulse's energy within an anti-Stokes absorption band. This creates an acoustic excitation in the fiber which retains the information content of the data pulses. In the retrieval process at a later time, a read pulse with the same direction as the write pulse depletes the acoustic wave and releases the data from the fiber, propagating in the same direction as the original data pulses. With this method 2 ns long data pulses have been stored up to 12 ns in a highly-nonlinear fiber (HNLF) of 5 m length. Thus, this is also a serious technique to facilitate light storage. In contrast to the EIT in Bose-Einstein condensates, the light storage by SBS can be implemented at room temperatures and works at any wavelength where the fiber is transparent. However, extremely high optical powers (a peak power of approximately 100 W for the write and read pulses) are required for this and the storage time is restricted by the lifetime of the acoustic excitation (e.g. 3.4 ns for pure silica [16]).

2.4.6 Wavelength-Conversion and Dispersion

During the last few years slow- and fast-light systems have been investigated which combine a change of the carrier wavelength of the optical signal pulse with a strong dispersion of a waveguide. This method contains and provides both, an exploitation of a material dispersion as well as a fine and coarse tuning. With this wavelength-conversion-dispersion method the highest

2.4 Slow- and Fast-Light Methods

time delays of 1.83 µs were achieved so far [76]. This was presented by a wavelength-transparent, all-optical delay of 10 Gbit/s signals using precise parametric dispersion engineering and triple wavelength-conversion in a HNLF [76]. With the same approach a 40 Gbit/s data channel was delayed by 1.56 µs [77].

In principle, the method is simply based on the frequency-dependent group index and group velocity of dispersive materials and not on gain or absorption spectra. If the center wavelength of an optical pulse is changed its v_g is changed as well due to the GVD. Similarly to the functionality of the waveguide-loops, this group velocity change together with the waveguide length L leads to a tunable group delay of the pulses which can be very high in materials with a high dispersion. The time delay ΔT of such systems can be defined as [4, p. 52]:

$$\Delta T = L\, D_{GVD}\, \Delta \lambda, \tag{2.23}$$

where D_{GVD} denotes the GVD parameter of the waveguide at the incident pulse carrier wavelength and $\Delta\lambda$ is the wavelength shift of the dispersion device. The wavelength-conversion in an optical fiber can be implemented by FWM, fiber-optic parametric processes or in periodically poled lithium niobate (PPLN) waveguides for instance [73–75, 78]. At the end of the delay line the output pulse wavelength λ_{out} has to be converted back to the initial carrier wavelength λ_{in}.

The most important advantage of the conversion-dispersion method compared to the resonance methods is that large amounts of controllable delays for data rates exceeding the 10 Gbit/s range with minimal pulse broadening at transmission wavelengths of 1550 nm can be achieved [4, 44]. For example, fractional delays of up to 1200 have been demonstrated for 3.5 ps pulses [98].

Although the conceptional idea of the wavelength-conversion-dispersion method is very simple its practical realization is complex and difficult. In most cases, many expensive components (e.g. tunable lasers and tunable filters) are necessary which need to be adapted to each other, in particular for every change of the time delay. Therefore, one crucial limiting factor is the system reconfiguration time which depends on the tuning speed of filters or pump laser frequencies [4, p. 54]. Additionally, the waveguide dispersion leads also to a temporal pulse broadening which degrades the time delay of high bit rate signals significantly. Hence, a sophisticated dispersion compensation has to be implemented. This further increases the complexity of the systems. For example, 185 km of SSMF were essential to compensate the pulse broadening of 105 ns delayed 40 Gbit/s and 80 Gbit/s signals [74]. Such a long fiber induces a large signal attenuation. Therefore, these systems are very power-consuming. At last, there are eligible discussions if the conversion-dispersion systems are really transparent for the pulses, i.e. if λ_{out} is related to λ_{in} [79–81].

2.4.7 Quasi-Light Storage

Recently, a new method to store single light pulses or pulse trains was found that exploits the coherence between the time and spectral representation of a light signal. It works like a 'real' optical memory, i.e. the input data signal remains stored as long as it will be released by a read signal. In an initial investigation delays of up to approximately 38 ns with several 5 bit patterns with bit durations of 500 ps were achieved [82].

The method works in principle by multiplying an input pulse with a frequency comb in the frequency domain which leads, according to the Fourier transform, to a pulse train with equidistant copies of the input pulse in the time domain. One of these copies is then extracted by a switch. It corresponds to a time-delayed replica of the input pulse. Thus, this method is called QLS.

In the experiment the filtering operation of the input pulse spectrum was implemented by SBS in a SSMF [82]. Since the minimum Brillouin bandwidth is limited to approximately 10 MHz [99], this technique provides maximum attainable storage times of 100 ns. Thereby, the number of stored bits is increased if higher bit rates are used. With rates of 2 Gbit/s and 10 Gbit/s for instance, maximum storage capacities of 200 bit and 1000 bit, respectively, are obtainable.

The significant advantages of the QLS method are that it only uses standard components of telecommunications and that the time delay can be tuned in both, the fine and the coarse range. Furthermore, high time delays of bursts can be achieved with tolerable optical powers and minimum signal distortions. In general, this method is independent from the bit rate, the modulation format and the wavelength of the signals.

2.5 Applications of Slow- and Fast-Light

The key feature of the slow- and fast-light effect is that a tunable optical delay line can be implemented. This enables a wide range of potential practical applications in many different fields. An overview is given in Figure 2.6.

The main applications can be found in optical communications and optical signal processing as the slow- and fast-light effect is meant to build up all-optical packet switches and routers which are essential to meet the demands of high bandwidth transmissions, low power consumption and physical space in future all-optical packet-switched networks. There, the delay line can be used primarily for optical buffers and memories [12, 50]. A very promising approach is to apply optical buffers for optical packet synchronization, multiplexing and equalization.

In addition, the effect is also very influential in the fields of physics, quantum optics and radio-frequency/microwave photonics. Applications such as time-resolved spectroscopy, nonlinear optics, quantum information processing and phased-array structures can be facilitated and drastically simplified or improved [12, 50, 100]. In the far future slow- and fast-light may open the door to all-optical computing. In the following paragraphs some of these most important

2.5 Applications of Slow- and Fast-Light

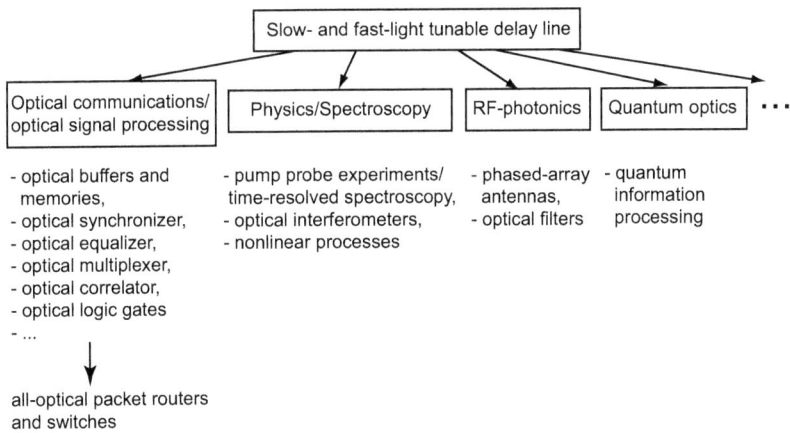

Figure 2.6: Applications of the slow- and fast-light effect.

potential applications are described.

2.5.1 Optical Buffers and Memories in Packed-Switched Networks

The most desirable application of the slow- and fast-light effect is the realization of all-optical packet switches and routers. Packet switching is a method of communication where the information is divided into blocks of various lengths addressed by a header [50]. These blocks are called packets. Inside the packet-switched network, such as the Internet, the packets are switched through different transmission paths by routers. They usually connect many networks and decide, according to the header, which path the packets take from its source to its destination. This also allows many users to share the same data path. The router can dynamically switch the data from any input port to any of the output ports [101]. A problem occurs if two packets arrive simultaneously at the router. This leads to a data packet contention and collision, as shown in Figure 2.7a. Thus, a temporary storage of one packet is necessary while the other one is processed (Figure 2.7b), i.e. its address is identified and the switching/routing is performed [100]. The stored packet is released when the first packet has left the router (Figure 2.7c).

In today's networks the data packets are transmitted in optical format and routed and switched in electronic format [50]. Therefore, the optical data stream is converted into the electrical domain, there it is buffered, proceeded and routed, and then it is reconverted into the optical domain. These optical-electrical-optical (OEO) conversion is just the bottleneck of the networks because it strongly limits the information bandwidth and consumes a lot of power and space [50, 100]. However, nowadays there is an increasing demand on transmission bandwidth and the data load doubles roughly each year [102]. These high bandwidths or frequencies accompany with inherent problems. First, the size of the electronic devices comes

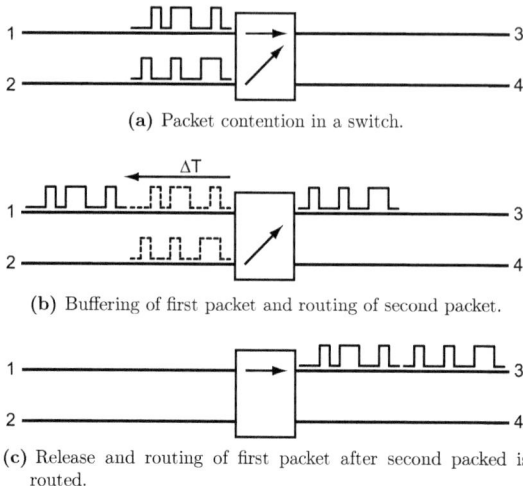

(a) Packet contention in a switch.

(b) Buffering of first packet and routing of second packet.

(c) Release and routing of first packet after second packed is routed.

Figure 2.7: Packet contention resolution in a switch.

into those of the wavelength ranges. Thus, every small component or conductor acts like an antenna which leads to strong signal interferences and losses. Therefore, the electrical data cannot be transmitted without suffering significant attenuations and distortions over long distances. Furthermore, the networks and hence, the physical size of the switches and routers is increased with the increasing transmission capacity. For these reasons, the electronic routers and switches are highly undesirable in the future whereas the all-optical counterparts can potentially eliminate this bottleneck [50].

All-optical networks are able to transmit more than 100 channels with data rates of 10 Gbit/s or 40 Gbit/s [102] and a large number of switches can be combined into a matrix on a small footprint [100]. While most functions of such an all-optical switch or router have already been demonstrated previously [10, 11, 103], the all-optical intermediate storage of the signals is still the missing key component. Such an all-optical buffer can be defined as follows [50]:

1. The data stream is completely optically. No OEO conversion is proceeded.

2. The buffer stores the signal for a time ΔT only with limited low distortions, attenuations or impairments.

3. The delay time ΔT is variable and externally controllable.

Conventional used devices which consist of sets of fixed fiber lengths, such as waveguide-loops as explained in Subsection 2.4.1, only provide fixed time delays. On the one hand, this leads immediately to a misalignment of the data streams and on the other hand, this constrains an easy adaption to different incoming data rates. Hence, these delay lines are not suitable for an efficient and flexible all-optical buffering. By contrast, a slow-light buffer fulfills the properties stated above and enables an accurate adjustment of the time delay and a precise allocation of each signal into a specific time slot with a dynamic adaptivity of the bit rate [4, pp. 324]. Therefore, such an optical buffer can be used for many more applications in optical communications [4, Chapter 16]. For example, an optical synchronization and multiplexing of multiple data channels on various fibers is utilizable.

However, if data streams with high bit rates are buffered the slow-light element may degrade the data quality due to the induced pattern dependence of the signal distortions. This problem can be solved by detuning the slow-light devices away from the signal carrier frequency [104] and adapting the adequate different modulation formats [105, 106]. Thus, all-optical buffers with the above-mentioned properties have been shown by using SBS for instance. Therefore, a 10 Gbit/s non-return-to-zero (NRZ) differential phase-shift-keyed (DPSK) signal has been transmitted error-free and continuously delayed by 42 ps [107] and 81.5 ps [108].

For example, an efficient two-by-one optical-time-division-multiplexer for three different input data streams at 2.5 Gbit/s, 2.67 Gbit/s and 5 Gbit/s was realized [109]. Thereby, the time slot of one signal path was manipulated relatively to the other by as much as 75 ps at a BER of 10^{-9}. Furthermore, an independent continuously-controllable delay of up to 112 ps and an all-optical bit-level synchronization of three 2.5 Gbit/s data channels within a single SBS-based slow-light element has been shown [97, 110]. Also by using a 26 ns tunable all-optical delay line based on wavelength-conversion and inter-channel chromatic dispersion, optical data packets of 10 Gbit/s have been synchronized and time-multiplexed [111].

Another potential application of all-optical buffers is the optical equalization. Most slow- and fast-light techniques are based on gain or absorption resonances. Hence, such optical delay lines can be additionally used to equalize the power of different data channels. Furthermore, a compensation of fiber dispersive effects and a mitigation of impairments of intersymbol interferences is possible [4, p. 325]. At last, one of the most elementary operations in signal processing is the correlation that needs also a time delay. This can be applied by an optical buffer too [50].

2.5.2 Spectroscopy and Interferometry

In time-resolved spectroscopy a pump-probe configuration is often used for example to perform surface analysis on materials. Thereby, two short pulses with a time delay relative to each other are sent into the sample. The pump pulse changes the properties of the material, whereas the time delayed probe pulse interrogates these changes. The time delay is achieved by mechanical components which change the length of the optical path. Hence, the precise adjustment of the delay is very complicated and the systems are usually very bulky. A compact optical buffer, based on slow- and fast-light, provides such an accurate handling and makes the systems compact [50].

Slow-light methods might also be applied in interferometry [12]. If the length of an interferometer is increased its spectral resolution can be enhanced significantly. However, this would increase the size of the system dramatically. Another way to emulate a long propagation path is to slow down the signal over a fixed length. By placing a slow-light medium within the interferometer structure this could be realized. Therefore, the resolution can be increased by a factor equal to the group index of the slow-light medium or, if the resolution is maintained, a minimized structure of the spectroscopic interferometer can be achieved [112, 113]. An enhancement of the spectral resolution of a Fourier transform interferometer by a factor of approximately 100 for example has been demonstrated [114]. There, a tunable slow-light medium was inserted into one of the two fixed interferometer arms.

2.5.3 Nonlinear Processes

The generation of nonlinear optical processes usually requires a sufficient amount of high optical powers and a large distance or time for the interaction with the material. This makes the equipment for nonlinearities often very large and expensive. The slow-light effect enables both, the miniaturization of the waveguides (up to μm scale) and the use of inexpensive components to produce an efficient nonlinear interaction. The increased optical intensity due to slow-light leads to a higher strength of the nonlinear interaction [100].

This enhanced nonlinear interaction enables a further foreseeable key application for optical communication techniques, the regenerator. Such a device cleans up a signal from distortions and noise induced by the propagation over a long distance of fiber and amplifier sections. The regeneration is done by the feature of the nonlinear transfer function. Currently, this requires high optical powers and thus, such systems are very costly and electrical regenerators are preferred instead. In contrast to this, the slow-light effect, as leading to moderate powers and costs, could make an optical regenerator possible [4, 100]. Furthermore, the delay or acceleration of several optical channels can be used to improve the phase-matching condition for nonlinear optical effects. For example, an enhanced all-optical control of the phase-matching condition in FWM processes was demonstrated by using the Brillouin slow-light effect in optical fibers [115].

2.5.4 Phased-Array Antennas

The most potential application in the field of radio-frequency photonics are phased-array devices. Phased-array antennas are also called smart antennas. These devices consist of a group of antennas in which the relative phases of their feeding signals are varied in such a way, that the directional characteristic of the array shows a strong directivity in a certain direction. By changing the phase relations of the feeding signals the direction of maximum gain is reconfigurable. The use of phased-array antennas is a very promising opportunity to improve the capacity of radio systems without increasing the number of base stations. To provide many radio services via one station smart antennas have to be wideband which can only be achieved by an optical control. Therefore, the electrical frequency-dependent phase change between the several feeding signals is replaced by an optical wavelength-independent time delay via slow- and fast-light elements [50, 70, 116].

2.5.5 Quantum Information Processing

In quantum information processing the quantum state has to be stored for a sufficient time to enable quantum operations. The storage can be easily implemented by the slow-light effect [100]. It has been shown that light pulses can be stored for example in a Bose-Einstein condensate without loosing their quantum properties [117].

Moreover, by slowing down two pulses simultaneously in the same material system also quantum operations can be performed. If the pulses propagate slow and with equal velocities they can efficiently interact for a long time. This leads to a close correlation state between the interacting photons that builds the basis of a quantum processor [100].

2.6 Summary

In this chapter an introduction to the topic of slow- and fast-light was made. At first, different velocities of light were defined. It was shown that slow- and fast-light is based on a change of the group velocity of light pulses. Furthermore, it was stated that this does not affect the signal velocity and therefore, no violation of the theory of special relativity in respect of superluminal velocities occurs.

The change of the group velocity can be achieved by a change of the group index which is in turn, according to the KKR, a consequence of a material resonance. To evaluate the grade of delay and slow- and fast-light efficiency different figures of merit and metrics were introduced. Among these, particularly the absolute, fractional and effective time delay (storage capacity) as well as the broadening factor are used for the analysis throughout this work.

The main focus of this chapter was an overview of the methods and possible applications of slow- and fast-light. It was shown that many different methods and material systems can be used to produce an alteration of the group velocity. These techniques can be separated in different categories and have different advantages. However, slow- and fast-light methods in optical fibers are of special interest because they can be included seamlessly into existing systems. Therefore, SBS, which exploits a strong material resonance for the group velocity alteration, is very promising for this. In general, the slow- and fast-light effect is attached to a wide range of potential applications for example in optical communications and optical signal processing but also in physics, quantum optics as well as in microwave photonics. Foreseen key applications in optical networks are optical delay lines or buffers for an all-optical synchronization, multiplexing and equalization for instance.

3 Slow- and Fast-Light Based on Stimulated Brillouin Scattering

SBS is the crucial effect to produce slow- and fast-light in this work. Hence, the generation of the slow- and fast-light effect based on SBS is described in this chapter. The fundamentals of the SBS itself will be explained at first. This includes the creation and properties of SBS, such as the threshold, the spectrum and the bandwidth. In Section 3.2 the mathematics, the experimental setup and the operation of a basic SBS-based slow- and fast-light system will be discussed. In the third section of this chapter a summary of the properties of standard SBS-based slow- and fast-light systems is given.

3.1 Brillouin Scattering

Brillouin scattering is named after the French physicist Leon Brillouin who investigated the interaction between light waves and acoustic waves in the nineteen-twenties [24, p. 269]. SBS is a nonlinear effect with a low threshold and it is the result of an interaction between an incident light wave and the material, such as an optical fiber. In general, the generation of nonlinear effects needs high intensities of light which have been available since the development of high efficient laser sources and fiber technologies. For this reason, SBS in optical fibers has not become of practical interest until the nineteen-sixties.

To transmit an optical signal over a long distance the optical input power could simply be increased to compensate the fiber attenuation. However, since the Brillouin scattering occurs at a few milliwatts of pump power in an optical fiber, the maximum transferable optical power is strongly limited. Hence, this effect generally disturbs the signal transmission in optical communication systems. Therefore, it is normally attempted to suppress SBS in optical components and networks [118–124].

However, during the last decades many applications which exploit and benefit from the properties of SBS have been investigated [125]. Besides slow- and fast-light, Brillouin scattering was used to implement fiber lasers [126–130], fiber amplifiers [131–135], optical fiber sensors [136–139] and tunable optical filters [140–142] for instance. Thus, it is possible to facilitate demodulator operations and channel selection in densely packed wavelength division multiplexing (WDM) systems [143–145], high resolution spectroscopy [146–149], millimeter wave generation in radio over fiber systems [150–154] as well as diagnostic tools for structural health monitoring to assessing civil structures [138, 155, 156].

3.1.1 Generation and Mode of Operation

SBS is based on an interaction of a sufficiently strong incident light wave (pump wave) with the optical medium, as shown in Figure 3.1. Density fluctuations inside the optical fiber or thermoelastic motions of the molecules cause a scattering of parts of the incident light wave into the backward direction. The backscattered wave is called Stokes-wave. This Stokes-wave superimposes with the pump wave which built up an electrical interference field (Figure 3.1b, #1). Due to the process of electrostriction[1] the interference pattern leads to a periodical density modulation of the medium (Figure 3.1b, #2). The density modulation can be seen as a modulation of the refractive index which acts like a Bragg grating. If the Bragg condition[2] is fulfilled more power of the pump wave is backscattered (Figure 3.1b, #3) [158]. This leads to a stronger density modulation which in turn exponentially increases the Stokes-wave. The process continues and more and more optical power of the pump wave is backscattered and transferred to the Stokes-wave (Figure 3.1b, #4). The Brillouin scattering effect becomes stimulated if the reason for the density modulation is the pump wave itself and the pump power exceeds a certain threshold [24, 159]. As long as the optical pump power supply is sufficiently strong and can compensate the pump power decrease due to the creation of the acoustic wave, the SBS process is self-preserving [160].

The density modulation of the medium can also be called acoustic phonons or acoustic wave, because it propagates with the speed of sound into the direction of the pump wave. Due to the relative velocity between the pump and the acoustic waves the frequency or wavelength of the Stokes-wave is shifted according to the Doppler effect[3]. This frequency or wavelength shift is called *Brillouin shift* [24, 159, 160].

The SBS can be seen as a classical coupled interaction of three waves. The scattering process requires that the momentum conservation law and energy conservation law have to be fulfilled for the interaction of the three waves [159]. Thus, in an optical fiber the pump frequency f_p, the Stokes frequency f_s and the acoustic frequency f_a are related by [24]:

$$f_s = f_p - f_a. \tag{3.1}$$

It can be seen from Equation (3.1) that the frequency of the Stokes-wave mediated by the acoustic wave is smaller than the frequency of the pump wave. Therefore, the frequency of the acoustic wave f_a equals the Brillouin shift f_B. It can be determined by [24]:

$$f_a = f_B = \frac{2v_a n}{\lambda_p}, \tag{3.2}$$

---

[1]Electrostriction describes the deformation of the material or the change of the density of the medium in the presence of an applied electric field [18], [24, p. 274].
[2]The Bragg condition is fulfilled if the phases of all parts of scattered or reflected waves are equal and interfere constructively. Therefore, the optical retardation between the reflected wave components has to be a multiple of their wavelength [157, p. 572]. In an optical fiber this depends on the grating period.
[3]The Doppler effect describes the change of a measured or an observed frequency of waves which is based on a relative motion between the source and the receiver [161]. In the case of SBS the pump wave has a relative motion or velocity in relation to the acoustic wave.

3.1 Brillouin Scattering

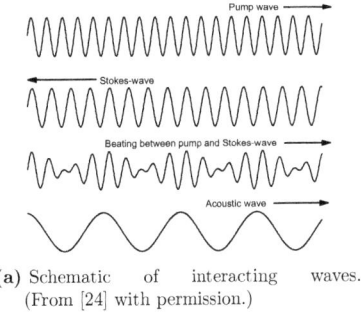

(a) Schematic of interacting waves. (From [24] with permission.)

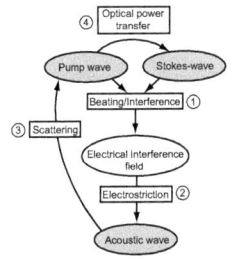

(b) Schematic mode of operation.

Figure 3.1: Generation of SBS.

where v_a is the speed of sound in the material, n is the refractive index of the medium and λ_p is the pump wavelength. Taking for example the following standard values for fused silica: $v_a = 5.96\,\text{km/s}$, $n = 1.44$ and $\lambda_p = 1550\,\text{nm}$ [162], the Brillouin frequency shift becomes $f_B \approx 11\,\text{GHz}$. Besides the variables of Equation (3.2), the Brillouin shift also depends on further parameters, such as the fiber type, the doping of the fiber core, the mechanical stress of the fiber or the ambient temperature [151]. For example, the temperature- and mechanical stress-dependent Brillouin shift change can be $1.36\,\text{MHz/°C}$ and $594.1\,\text{MHz/\%}_{\text{elongation}}$ in SSMF [163]. For the fibers used in this work a Brillouin shift of $10.87\,\text{GHz}$ was determined. The measurement of f_B is explained in Section A.1.

The Stokes-wave, which is beating with the pump wave, can also be produced by launching a signal wave at the opposite end of the fiber [138]. Therefore, the signal wave has to be frequency-shifted by the Brillouin shift f_B. If the pump power is below a certain threshold it only creates a Brillouin-gain region and Brillouin-loss region inside the fiber [164]. Thereby, the gain (Stokes band) is frequency downshifted and the loss is frequency upshifted (anti-Stokes band) by f_B. Within these regions the counterpropagating signal wave can be amplified (power transfer from the pump to the Stokes-wave) or attenuated (power transfer from the Stokes-wave to the pump wave) [165]. The required pump power to preserve the SBS process is decreased in this case, because the launched signal wave (Stokes-wave) delivers additional power to the scattering effect [160].

3.1.2 Brillouin Threshold and Brillouin Gain Coefficient

SBS occurs at a very low pump power threshold. The knowledge of this threshold is essential to configure optical networks and to control systems based on SBS. If the pump power exceeds the threshold an exponential growing Stokes-wave is created by the noise in the fiber. The SBS reaches a saturation where the pump power is depleted and the transmitted power is restricted. For this pump depletion a few milliwatts are sufficient in optical fibers of a few tens of kilometers. Below the threshold the Stokes-wave is very weak and the SBS process creates

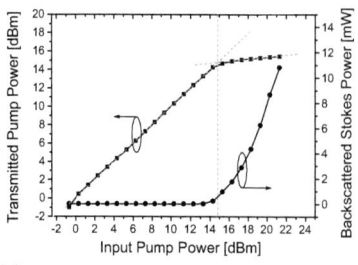
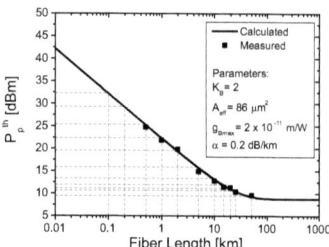

(a) Brillouin threshold measurement of a SSMF with a length of 5 km. The fiber attenuation was neglected.

(b) Calculated and measured Brillouin threshold as a function of the fiber length.

Figure 3.2: Calculated and measured Brillouin threshold.

only a Stokes-gain and an anti-Stokes-absorption region for counterpropagating signal waves. The Brillouin threshold can be variously defined [24, 166], for instance as:

1. the optical input pump power which is equal to the backscattered Stokes power [167],

2. the optical input pump power at which the backscattered Stokes power is equal to the transmitted pump power [168],

3. the optical input pump power at which the backscattered Stokes power begins to increase strongly or the transmitted pump power begins to deplete [169, 170],

4. the optical input pump power at which the backscattered Stokes power is equal to 1 % of the input pump power [171] or

5. the optical input pump power at which the backscattered Stokes power is equal to the backscattered Rayleigh power [172].

For practical purposes the threshold power can also be determined by the maximum input pump power which is transmittable before the SBS effects become detectable or in some sense significant [162]. This condition correlates best with the third Brillouin threshold definition which is therefore, used in this book.

The practical measurement of the Brillouin threshold is explained in detail in Section A.2. Figure 3.2a shows measurement results and the Brillouin threshold determination for a 5 km SSMF. As can be seen from the diagram, the SBS activity starts at an Brillouin threshold of approximately 14.8 dBm which corresponds to approximately 30 mW. Above this power value the backscattered Stokes-wave rapidly increases whereas the transmitted pump power reaches the saturation regime. The measurement diagrams of the other used fibers are shown in Figure A.4 in Section A.2.

3.1 Brillouin Scattering

The Brillouin threshold P_p^{th} is commonly calculated by [24, 159, 168]:

$$P_p^{th} = 21 \frac{K_B A_{eff}}{g_{Bmax} L_{eff}}, \qquad (3.3)$$

where K_B is a polarization factor, g_{Bmax} is the maximum value of the Brillouin gain coefficient in the line center and L_{eff} and A_{eff} denotes the effective length and effective area of the fiber. Although Equation (3.3) is widely used there are some recommendations to adapt the constant factor from 21 to 19 [24, 164, 166, 173]:

$$P_p^{th} = 19 \frac{K_B A_{eff}}{g_{Bmax} L_{eff}}. \qquad (3.4)$$

The polarization factor K_B takes the birefringence of the fiber into consideration and can take on values between 1 and 2 [174, 175]. The exact value depends on the polarization relation between the pump and the Stokes-wave or counterpropagating signal wave. For an optimal interaction between the pump and the Stokes-waves, both waves have to be co-polarized along the entire length of the fiber. In this case K_B is minimized which is valid for polarization maintaining fibers for instance. In conventional fibers, such as SSMF, the polarization states of both counterpropagating waves differ at various places in the fiber. Hence, the strength of the interaction is decreased and $K_B = 2$ is generally used as an average value [160].

The effective length L_{eff} depends on the absolute length L and the attenuation constant $\alpha_{km^{-1}}$ in [km^{-1}] of the fiber [24, p. 99]:

$$L_{eff} = \frac{1 - e^{-\alpha_{km^{-1}} L}}{\alpha_{km^{-1}}}. \qquad (3.5)$$

It follows that the effective length increases for longer fibers, but approximates a maximum value. Hence, the Brillouin threshold is decreased for long fibers and reaches a minimum value which is defined by the attenuation constant of the fiber, as can be seen in Figure 3.2b. Commonly used SSMF have an attenuation constant of $\alpha_{dB/km} = 0.2$ dB/km which corresponds to $\alpha_{km^{-1}} \approx 0.046$ km^{-1}. Therefore, the minimum Brillouin threshold for long fibers is approximately 8.76 dBm, if $g_{Bmax} = 2 \times 10^{-11}$ m/W, $K_B = 2$ and $A_{eff} = 86\,\mu m^2$ are assumed. This very low threshold value indicates that the SBS reaches the saturation regime and the pump depletes already very fast at low pump powers in long optical fibers.

For the measurements in this work SSMF with different lengths were used. All these fibers originated from the same manufacturing process and were spooled similarly. Hence, it can be assumed that their material parameters regarding the SBS, especially the Brillouin gain coefficient, are similar. Therefore, the diagram in Figure 3.2b also shows that the measured Brillouin thresholds of the used fibers are in good agreement with the theoretical prediction. For example, the threshold value of a SSMF with an absolute and effective length of 20 km and 13.070 km, respectively, is $P_p^{th} \approx 11$ dBm or approximately 13 mW. The calculated and measured Brillouin thresholds of all used SSMF are shown in Table A.2.

The peak value of the Brillouin gain coefficient g_{Bmax} is the maximum value of the Brillouin gain distribution at the frequency of $f_p - f_B$ if the polarization relation between the pump and the Stokes-wave is neglected [159]. The coefficient depends on different material properties of the fiber which are combined to the elasto-optic figure of merit M_2. The maximum SBS gain coefficient can be described by [24]:

$$g_{Bmax} = \frac{\pi f_B^2 M_2}{c \, \Delta f_B \, 2n}. \tag{3.6}$$

As can be seen from Equation (3.6), the peak value of the Brillouin gain coefficient depends on the Brillouin shift f_B and the Brillouin FWHM bandwidth Δf_B and therefore, also on the mechanical stress, the temperature and the dopant concentration of the fiber. Typical values of g_{Bmax} lie in the range between 1×10^{-11} m/W and 5×10^{-11} m/W in SSMF at a pump wavelength of 1550 nm [158]. For fused silica $M_2 = 1.51 \times 10^{-15}$ s^3/kg [167]. Thus, a Brillouin shift of 11 GHz and a Brillouin bandwidth of 35 MHz for instance yields a maximum Brillouin gain coefficient of $g_{Bmax} \approx 1.92 \times 10^{-11}$ m/W. However, in practice the material properties of the fiber are often not provided by the manufacturers. Hence, the exact value of M_2 is unknown. Thus, g_{Bmax} can be determined by using Equation (3.4) with a measured Brillouin threshold power. For example, the maximum SBS gain coefficient g_{Bmax} was determined to be approximately 1.92×10^{-11} m/W for a 20 km SSMF with the measured threshold of approximately 13 mW, an effective length of 13 070 m, an effective area of 86 μm^2 and with $K_B = 2$. This experimentally determined value of g_{Bmax} agrees well with the theoretical value.

For the practical use of the maximum Brillouin gain coefficient the polarization relation between the pump and the Stokes-wave cannot be neglected. If the polarization states of both waves are not perfectly matched, g_{Bmax} will be reduced by K_B. At the same time the Brillouin threshold is increased. The polarization-dependent Brillouin gain coefficient g_B becomes [160, 162]:

$$g_B = \frac{g_{Bmax}}{K_B} \tag{3.7}$$

By considering the polarization dependence it follows for the measured Brillouin gain coefficient that $g_B = 0.96 \times 10^{-11}$ m/W. From the Brillouin threshold measurements of all used fibers an average gain coefficient value of $g_B \approx 1 \times 10^{-11}$ m/W was determined. Henceforth, this value is used throughout this work. An overview of all relevant fiber parameters is shown in Section A.4.

3.1.3 Intensity Equations and Brillouin Amplifier Gain

The interaction between the pump and the Stokes-wave along the fiber can be described by a system of two coupled differential equations for the pump and Stokes-intensities I_p and I_s [24, 159]:

3.1 Brillouin Scattering

$$\frac{dI_s}{dz} = -g_B I_p I_s + \alpha I_s, \quad (3.8)$$

$$\frac{dI_p}{dz} = -g_B I_p I_s - \alpha I_p, \quad (3.9)$$

where α is the fiber attenuation. Since the frequency shift between the pump wave and the Stokes-wave is rather small, α and g_B are equal for both waves. Additionally, Equation (3.8) and Equation (3.9) are only valid for approximately monochromatic waves [24]. From Equation (3.8) it can be derived that the backscattered Stokes-wave increases exponentially with negative values of z as long as the product of I_p and I_s is greater than the fiber attenuation. Simultaneously, due to the SBS process which can be deduced from Equation (3.9) the pump wave is decreased with positive values of z by the fiber attenuation and the power transfer to the Stokes-wave. The conditions of the intensities along the fiber with the length L is illustrated in Figure 3.3.

If the pump intensity is small and the pump power is below the threshold the pump depletion and hence, the power transfer from the pump to the Stokes-wave can be neglected [24]. Thus, the first term on the right hand side of Equation (3.9) vanishes and the pump intensity depends on the fiber attenuation only:

$$I_p(z) = I_p(0) e^{-\alpha z}. \quad (3.10)$$

Accordingly, Equation (3.8) can be rewritten [24]:

$$\frac{dI_s}{dz} = [-g_B I_p(z) + \alpha] I_s. \quad (3.11)$$

By introducing Equation (3.10) into this differential equation, the Stokes intensity at the fiber input ($z = 0$) can be solved to [24, 159]:

$$I_s(0) = I_s(L)\, e^{g - \alpha L}, \quad (3.12)$$

with

$$g = \frac{g_B P_p L_{eff}}{A_{eff}}. \quad (3.13)$$

The gain factor G of the Brillouin amplifier is the ratio between the Stokes intensity I_s or power P_s at $z = 0$ (output for the signal or Stokes-wave) and the Stokes intensity or power at $z = L$

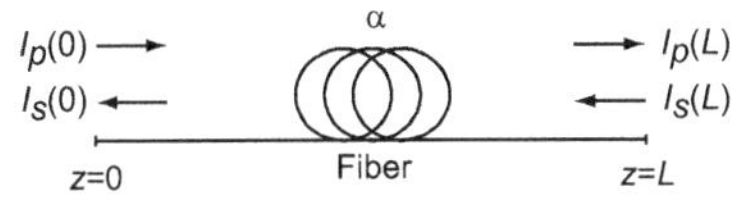

Figure 3.3: Propagation conditions of the pump and Stokes-wave along the fiber.

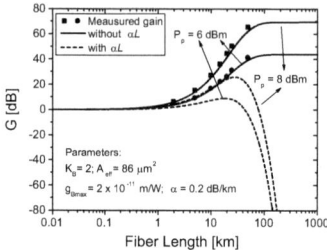

Figure 3.4: Calculated and measured Brillouin gain as a function of the fiber length for two constant pump powers.

(input for the signal or Stokes-wave) [24]:

$$G = \frac{I_s(0)}{I_s(L)} = \frac{P_s(0)}{P_s(L)} = e^{g-\alpha L}. \qquad (3.14)$$

According to Equation (3.14), the logarithmic relation leads to the logarithmic gain factor G_{dB}:

$$G_{dB} = 10 \log \left(e^{g-\alpha L} \right) \approx 4.343 \left(g - \alpha L \right). \qquad (3.15)$$

By including Equation (3.4) into Equation (3.15) and if the fiber attenuation is neglected, the Brillouin gain threshold G_{dB}^{th} can be determined:

$$G_{dB}^{th} = 4.343 \times 19 \approx 82 \, \text{dB}. \qquad (3.16)$$

Thus, the Brillouin gain threshold in SSMF is constant and independent from the fiber and environmental parameters in contrast to the amplifier gain factor.

Equation (3.12) and Equation (3.14) imply that an initiating counterpropagting signal or Stokes-wave with the correct frequency shift with respect to the pump wave is necessary at the end ($z = L$) of the fiber [24]. The first part of the exponent in Equation (3.12) is the Brillouin gain g which is responsible for the exponential amplification of the backscattered Stokes-wave induced by the pump power P_p. The second term of the exponent in Equation (3.12) leads to an exponential attenuation of the Stokes-wave along the fiber.

The Brillouin amplifier gain as a function of the fiber length for two constant pump powers P_p is shown in Figure 3.4. The solid lines represent the calculation without the fiber attenuation. In general the gain increases with the fiber length for a constant pump power until a maximum value is reached and the Brillouin amplifier begins to saturate. Thereby, higher pump powers lead to higher achievable gains. This behavior was confirmed by an experimental determination of the gain for the used SSMF at the two given pump powers. For this purpose the gain was defined as the ratio of the measured Stokes power with and without SBS at the output of the Brillouin amplifier. Therefore, the fiber attenuation was not considered in this case. As shown by the symbols in the diagram, the measured gain values agree well with the calculated gain

3.1 Brillouin Scattering

progress.

In practice, the term αL cannot be neglected, as can be seen from the dashed lines in Figure 3.4. As long as the amplification is greater than the attenuation, the intensity of the Stokes-wave is increased up to a maximum value. At the same time the entire fiber attenuation grows further with the fiber length. If the impact of the attenuation is greater than the amplification the efficiency of the SBS process is decreased and the signal wave will be strongly attenuated.

3.1.4 Spectral Characteristics

In Subsection 3.1.2 the peak value of the Brillouin gain coefficient was introduced by Equation (3.6) and Equation (3.7). These equations are only valid for the maximum of g_B exactly at the frequency of the acoustic wave, i.e. at the frequency shift $f_p - f_B$ or the line center of the gain. In general, the Brillouin gain is a function of the frequency and forms a gain distribution with a narrow bandwidth. For quasi-monochromatic pump waves with a narrow linewidth the shape of the natural Brillouin spectrum can be approximated by a Lorentzian function [24, 159]:

$$g_B(f) = \frac{(\Delta f_B/2)^2}{(f - f_B)^2 + (\Delta f_B/2)^2} g_B(f_B). \tag{3.17}$$

Thereby, the Lorentzian shape is related to the exponential attenuation of the acoustic waves [158–160]. The natural FWHM bandwidth Δf_B depends on the lifetime of the acoustic phonons T_a which depends on the acoustic attenuation constant α_a. Therefore, the Brillouin bandwidth can be calculated by [176]:

$$\Delta f_B = \frac{1}{T_a} = \frac{\alpha_a v_a}{2\pi}. \tag{3.18}$$

Generally, the acoustic lifetime is in the range of a few ns. For silica for instance it is approximately 10 ns [4, p. 173] which corresponds to a bandwidth of 100 MHz. In SSMF at pump wavelengths of 1550 nm the Brillouin bandwidth lies typically in the range of 10 MHz to 50 MHz [24, 158]. Furthermore, the bandwidth directly changes with the pump power and Brillouin amplifier gain, respectively. With increasing input powers Δf_B is narrowed. It reaches a constant value and can be as narrow as 10 MHz if the pump power exceeds the threshold where the system operates in the pump-depleted regime [99].

If the linewidth of the pump waves is increased the Brillouin spectrum increasingly differs from a Lorentzian shape and merges into a Gaussian distribution [24]. Generally, the final Brillouin spectrum is a result of the convolution between the natural Brillouin bandwidth Δf_B and the linewidth of the pump source Δf_p [24, 166, 168]:

$$\Delta f_B \otimes \Delta f_p. \tag{3.19}$$

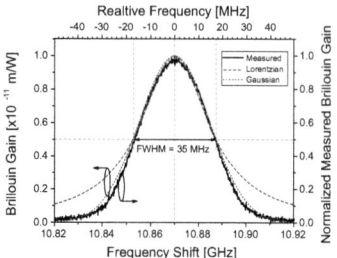

Figure 3.5: Measured Brillouin gain spectrum in comparison to a Lorentzian and Gaussian distribution.

For a Gaussian distribution Equation (3.19) becomes [24, 166]:

$$\sqrt{\Delta f_B^2 + \Delta f_p^2}, \qquad (3.20)$$

and for a Lorentzian distribution [24, 166]:

$$\Delta f_B + \Delta f_p. \qquad (3.21)$$

Due to this behavior the Brillouin bandwidth becomes wider if the pump waves have a broad linewidth. Simultaneously, the Brillouin gain coefficient g_{Bmax} is decreased whereas the Brillouin threshold P_p^{th} is increased by the same order of magnitude of the broadening. Hence, Equation (3.6) has to be extended by the factor[4] [24, 166]:

$$\frac{\Delta f_B}{\Delta f_B \otimes \Delta f_p}. \qquad (3.22)$$

This means that g_{Bmax} constitutes only 87.5 % if for example a distributed feedback (DFB) laser diode (LD) with a typical linewidth of 5 MHz is used as pump source and a Lorentzian gain profile with a bandwidth of 35 MHz is assumed. Additonally, the Brillouin threshold is increased by 14 %. Hence, the Brillouin interaction is decreased. This feature provides an opportunity to model the Brillouin spectrum by designing the spectrum of the pump wave, as will be described in the following chapters.

For the optical fibers used in this work a natural Brillouin bandwidth of approximately 35 MHz was determined. The measurement of the Brillouin spectrum is explained in Section A.3. In Figure 3.5 the measured Brillouin spectrum can be seen in comparison to the theoretical Lorentzian and Gaussian distribution at the Brillouin shift. The diagram clearly shows the deviation of the measured spectrum from the theoretical Lorentzian shape. The reason for that can be attributed to the non-quasi-monochromatic pump wave of the DFB-LD. Hence, the measured gain distribution shows a rather Gaussian shape.

---

[4] At the same time Equation (3.3) and Equation (3.4) have to be extended by the reciprocal of this factor.

3.2 SBS-Based Slow- and Fast-Light

Slow- and fast-light via SBS was first demonstrated independently by Song et al. [13] and Okawachi et al. [14] in 2005. In Song's experiment 100 ns pulses were continuously delayed between -10 ns and $+30$ ns with a slope of $0.97\,\text{ns}/\text{dB}_{\text{gain}}$ in a SSMF. Okawachi's experiment shows a delay of 63 ns and 15 ns input pulses by 25 ns and 20 ns which corresponds to fractional delays of 0.4 and 1.3. Quite similar results were achieved in [15], where the delay of the 100 ns pulse was tuned from -14.4 ns to $+18.6$ ns in a SSMF of only 2 m in length. Due to the extremely short fiber extreme values of the group index and group velocity were observed. Compared to the normal conditions where $n_g = 1.46$ ($v_g \approx 2.05 \times 10^8$ m/s), at the highest positive time delay $n_g = 4.26$ ($v_g \approx 0.705 \times 10^8$ m/s), and at the highest negative time delay $n_g = -0.7$ ($v_g \approx -4.28 \times 10^8$ m/s). These results show that SBS enables a continuous tuning of the group velocity over a very large range with achievable fractional time delays > 1. Because of this ability, the SBS has become a very attractive candidate for slow- and fast-light systems.

In this section it will be shown how the slow- and fast-light effect works on the basis of SBS. At first, the mathematical background theory is described. After that, the principle experimental setup to produce SBS-based slow- and fast-light is shown and explained. At last, first measurement results of a system with a single natural gain are discussed.

3.2.1 Background Theory

As explained in Section 3.1, a pump wave creates a frequency-shifted Stokes-gain and an anti-Stokes-loss region if its power is below the Brillouin threshold. According to the KKR, as explained in Section 2.2 and shown in Figure 3.6, these gain and loss resonances lead to a phase change which in turn results in a change of the group index Δn_g. On the one hand, a counter propagating signal wave can be amplified or attenuated within the gain and the loss region. On the other hand, the associated positive or negative group index change additionally causes a lower or higher group velocity of the signal. Hence, the Brillouin gain yields slow-light whereas the Brillouin loss leads to fast-light.

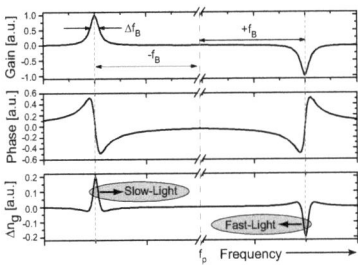

Figure 3.6: Stokes-gain and anti-Stokes-loss (top), phase distribution (middle), and group index change (bottom) due to SBS.

This behavior can again be theoretically explained by the system of two coupled differential equations. According to Equation (3.8) and Equation (3.9), the interaction between the pump field E_p and signal or Stokes-field E_s can be described as [177–179]:

$$\frac{dE_s}{dz} = -\left(\frac{g_B}{2A_{\text{eff}}}P_p\frac{\alpha_a}{2}\Delta k_R - \frac{\alpha}{2}\right)E_s + j\left(\frac{g_B}{2A_{\text{eff}}}P_p\frac{\alpha_a}{2}\Delta k_I\right)E_s, \qquad (3.23)$$

$$\frac{dE_p}{dz} = -\left(\frac{g_B}{2A_{\text{eff}}}P_s\frac{\alpha_a}{2}\Delta k_R + \frac{\alpha}{2}\right)E_p - j\left(\frac{g_B}{2A_{\text{eff}}}P_s\frac{\alpha_a}{2}\Delta k_I\right)E_p, \qquad (3.24)$$

where Δk_R and Δk_I are the real and imaginary parts of the complex phase mismatch term Δk:

$$\Delta k = \Delta k_R + j\Delta k_I = \left(\frac{\alpha_a}{2} - j\frac{\omega - \omega_0}{v_a}\right)^{-1}. \qquad (3.25)$$

In the undepleted regime the pump power depends only on the fiber attenuation and is therefore $P_p(L) = P_p(0)L_{\text{eff}}$ [24]. Thus, with $g = g_B P_p L_{\text{eff}}/A_{\text{eff}}$ the solution of the differential equation for the Stokes-field at the fiber input yields:

$$E_s(0) = E_s(L)\exp\left(g\frac{\alpha_a}{4}\Delta k_R - \frac{\alpha L}{2} - jg\frac{\alpha_a}{4}\Delta k_I\right). \qquad (3.26)$$

The real part of the exponent in Equation (3.26) leads to a change of the amplitude of the Stokes field. According to Figure 3.3, there will be a Stokes amplification (g is positive) or anti-Stokes absorption (g is negative) along the fiber from $z = L$ to $z = 0$. Simultaneously, the fiber attenuation decreases the Stokes-wave. The imaginary part of the exponent in Equation (3.26) is responsible for the phase change which causes the change of the group index and induces the time delay due to SBS.

All these changes depend on the phase mismatch Δk. Via a conjugate-complex extension Equation (3.25) can be rewritten as [177, 180]:

$$\begin{aligned}\Delta k = \Delta k_R + j\Delta k_I &= \frac{\frac{v_a^2 \alpha_a}{2}}{\left(\frac{\Delta\omega_B}{2}\right)^2 + (\omega - \omega_0)^2} + j\frac{v_a(\omega - \omega_0)}{\left(\frac{\Delta\omega_B}{2}\right)^2 + (\omega - \omega_0)^2} \\ &= \frac{4v_a}{\Delta\omega_B}\left[\frac{\frac{1}{2}}{1 + 4\left(\frac{\omega - \omega_0}{\Delta\omega_B}\right)^2} + j\frac{\frac{\omega - \omega_0}{\Delta\omega_B}}{1 + 4\left(\frac{\omega - \omega_0}{\Delta\omega_B}\right)^2}\right],\end{aligned} \qquad (3.27)$$

with $\Delta\omega_B = 2\pi\Delta f_B$ as angular Brillouin FWHM bandwidth. Hence, the real part of the exponent in Equation (3.26) without the fiber attenuation becomes:

$$g\frac{\alpha_a}{4}\Delta k_R = g\frac{\frac{1}{2}}{1 + 4\left(\frac{\omega - \omega_0}{\Delta\omega_B}\right)^2}. \qquad (3.28)$$

This equation describes the frequency-distribution of the Brillouin gain or loss and therefore,

3.2 SBS-Based Slow- and Fast-Light

the Brillouin amplification or absorption. It shows a Lorentzian shape with the Brillouin bandwidth $\Delta\omega_B$. Since, Equation (3.28) is derived from field values it is only the half of the distribution described by Equation (3.17) in Subsection 3.1.4. However, in practice the gain and loss distribution has to be related to power or intensity values. Thus, $g_B(f) = 2g\Delta k_R \alpha_a/4 = g\Delta k_R \alpha_a/2$, because $|E_s|^2 \propto P_s$ or I_s. The Lorentzian-shaped gain and loss distribution is shown in the upper diagram of Figure 3.6.

Together with Equation (3.27) the imaginary part of the exponent in Equation (3.26) becomes:

$$-g\frac{\alpha_a}{4}\Delta k_I = -g\frac{\frac{\omega-\omega_0}{\Delta\omega_B}}{1+4\left(\frac{\omega-\omega_0}{\Delta\omega_B}\right)^2}. \tag{3.29}$$

This equation describes the phase change of the Stokes wave which is shown in the middle diagram of Figure 3.6. The additive inverse of Equation (3.29) corresponds to the frequency distribution of the refractive index $g\Delta k_I \alpha_a/4 \propto n$. Hence, Equation (3.28) and Equation (3.29) are related by the KKR. The derivative of n with respect to the frequency yields the change of the group index Δn_g which in turn is proportional to the time delay due to SBS. The group index change is shown in the lower diagram of Figure 3.6.

The propagation of the signal pulse through the slow- and fast-light medium can generally be described by a transfer function $H(\omega)$ in the frequency domain. This transfer function can be deduced from Equation (3.26) as the relation of the pulse output and input fields or amplitudes [181]:

$$H(\omega) = \frac{E_s(0)}{E_s(L)} = |H(\omega)|e^{j\angle H(\omega)} = e^{jk(\omega)L}, \tag{3.30}$$

with $k(\omega)$ as the complex wave number which describes the propagation of the pulse along the fiber length L. The absolute value $|H(\omega)|$ corresponds to the amplitude and the argument $\angle H(\omega)$ characterizes the phase of the signal pulse. As compared with Equation (3.26), the influence of $k(\omega)$ on the pulse propagation corresponds to the impact of the complex phase mismatch term Δk. Hence, in the following only the complex wave number will be considered to describe the propagation of the pulse through the slow- and fast-light system.

Since the pulse consists of different frequencies around ω_0 the complex wave number includes many frequency components. Hence, $k(\omega)$ can be evolved into a Taylor series [24, 181]:

$$k(\omega) = \sum_{i=0}^{\infty} \frac{k_i(\omega-\omega_0)^i}{i!} = k_0 + k_1(\omega-\omega_0) + \frac{1}{2}k_2(\omega-\omega_0)^2 + \ldots, \tag{3.31}$$

with

$$k_i = \left[\frac{d^i k(\omega)}{d\omega^i}\right]_{\omega=\omega_0}. \tag{3.32}$$

In a vacuum and a dispersionless medium $k(\omega)$ consists of only the first term $k_0 = \omega/c$ and $k_0 = n\omega/c$, respectively. In contrast to this, a dispersive medium contains also the higher-order terms of $k(\omega)$ whereby the first derivative $k_1 = dk(\omega)/d\omega$ corresponds to the reciprocal group velocity v_g^{-1}. This leads to the transmission time of the signal pulse through the fiber. The second derivative $k_2 = d^2k(\omega)/d\omega^2 = d(v_g^{-1})/d\omega$ is the GVD which is one reason for the distortion of the pulses during the propagation [24]. The higher-order terms of the complex wave number can lead to further pulse distortions or changes in the pulse shape. If the third and higher-order terms of Equation (3.31) are zero and the second term is real and linear the pulse will only be amplified or attenuated, phase-shifted and delayed or accelerated, but the pulse experiences no distortions and its shape remains unchanged [181]. However, inside a real transmission medium the higher-order terms of Equation (3.31) exist and contributes to the pulse distortions.

Due to the SBS resonance the complex wave number of the pulse inside the medium is strongly manipulated. Therefore, it is composed of the linear term $k_0 = n\omega/c$ and a complex Brillouin-based wave number $k_{SBS}(\omega)$ which is related to the complex phase mismatch term Δk:

$$k(\omega) = \frac{n}{c}\omega + k_{SBS}(\omega). \tag{3.33}$$

As previously described, the first derivative of $k(\omega)$ leads to the reciprocal group velocity. Hence, it also reveals information about the group index change. Due to the linearity of the first term of Equation (3.33) its derivative is constant. Therefore, the group index change and the time delay depend only on the SBS effect.

In general, the pulse delay t_{del} can be defined as the difference between the transmission time in a vacuum $t_1 = L/c$ and the transmission time in the medium via group delay $t_2 = t_g = L/v_g = Lk_1$:

$$t_{del} = t_2 - t_1 = Lk_1 - \frac{L}{c} = L\left(k_1 - \frac{1}{c}\right) = L\left(\frac{dk(\omega)}{d\omega} - \frac{1}{c}\right). \tag{3.34}$$

According to Equation (3.26) and Equation (3.27) the complex wave number of a single natural Brillouin gain or loss can be written as [165, 181, 182]:

$$\begin{aligned}k(\omega) &= \frac{n}{c}\omega + \frac{g}{2L}\left[\frac{\gamma}{(\omega-\omega_0)+j\gamma}\right]\\ &= \frac{n}{c}\omega + \frac{g}{2L}\left[\frac{\gamma(\omega-\omega_0)}{(\omega-\omega_0)^2+\gamma^2} - j\frac{\gamma^2}{(\omega-\omega_0)^2+\gamma^2}\right],\end{aligned} \tag{3.35}$$

where $\gamma = \pi\,\Delta f_B$ is the angular half width at half maximum (HWHM) bandwidth of the Brillouin spectrum. Therefore, the complex transfer function becomes:

$$H(\omega) = \exp\left\{\frac{g}{2}\frac{\gamma^2}{(\omega-\omega_0)^2+\gamma^2} + j\left[\frac{n}{c}\omega L + \frac{g}{2}\frac{\gamma(\omega-\omega_0)}{(\omega-\omega_0)^2+\gamma^2}\right]\right\}. \tag{3.36}$$

The real part of the exponent of the complex transfer function, i.e. the imaginary part of the wave number, leads again to the amplitude change (amplification or attenuation) of the pulse

3.2 SBS-Based Slow- and Fast-Light

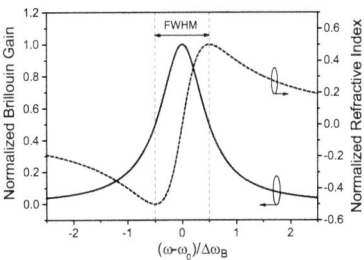

(a) Gain (solid) and refractive index (dashed) distribution.

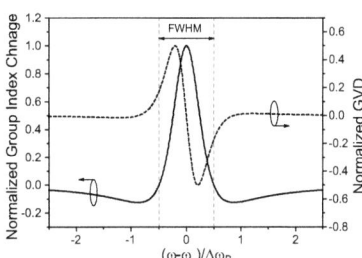

(b) Group index change (solid) and GVD (dashed).

Figure 3.7: Behavior of a Brillouin gain resonance. A Brillouin absorption resonance shows exactly the reverse behavior.

and corresponds to the expression in Equation (3.28). Thus, the frequency-distribution of the Brillouin gain $g_B(\omega) = 2\Re[jk(\omega)L]$ shows a Lorentzian shape, as shown by the solid line in Figure 3.7a.

The imaginary part of the exponent in Equation (3.36), i.e. the real part of Equation (3.35) results in the phase change of the signal pulse and hence, leads to a frequency-distribution of the refractive index. This is equal to the additive inverse of Equation (3.29) if the linear term is neglected. The refractive index exhibits a positive dispersion within the gain FWHM bandwidth, as it is shown by the dashed line in Figure 3.7a. Finally, the derivative $d\{\Im[jk(\omega)L] - n\omega L/c\}/d\omega \propto dn/d\omega$ leads to the group index change Δn_g, as can be seen in Figure 3.7b.

The first derivative of $k(\omega)$ yields:

$$\frac{dk(\omega)}{d\omega} = k_1 = \frac{1}{v_g} = \frac{n}{c} - \frac{g}{2L}\frac{\gamma}{[(\omega - \omega_0) + j\gamma]^2}, \tag{3.37}$$

with

$$\Re(k_1) = \frac{n}{c} + \frac{g}{2L}\frac{\gamma^3 - \gamma(\omega - \omega_0)^2}{\left[(\omega - \omega_0)^2 + \gamma^2\right]^2}. \tag{3.38}$$

Hence, by including Equation (3.38) into Equation (3.34) the pulse delay becomes:

$$t_{del} = \frac{L}{c}(n-1) + \frac{g}{2}\frac{\gamma^3 - \gamma(\omega - \omega_0)^2}{\left[(\omega - \omega_0)^2 + \gamma^2\right]^2}. \tag{3.39}$$

The first term of Equation (3.39) describes the transmission time of the pulse through the medium and is the difference between the transfer times with group velocity and speed of light in vacuum. The second part of Equation (3.39) is the additional pulse delay only caused by the Brillouin resonances.

The maximum pulse delay occurs in the line center of the Brillouin spectrum at $\omega = \omega_0$.

Thus, the maximum time delay induced by SBS with a single natural gain or loss is:

$$\Delta T = \frac{g}{2\gamma} = \frac{g}{2\pi\,\Delta f_B} = \frac{g_B P_p L_{eff}}{A_{eff}} \frac{1}{2\pi\,\Delta f_B} \approx 0.23 \frac{G_{dB}}{2\pi\,\Delta f_B}. \quad (3.40)$$

As can be seen from Equation (3.40) in the case of a Brillouin gain, where g has positive values, the time delay is positive which corresponds to slow-light. In the case of a Brillouin loss g is negative. This leads to a negative time delay. The group velocity of the signal pulse is advanced and fast-light is provided. Thereby, the time delay is directly proportional to the applied gain or loss and pump powers as well as inversely proportional to the Brillouin bandwidth which depends on the pump linewidth. Hence, the time delay is controllable by adjusting P_p and Δf_B. The other parameters in Equation (3.40) are constant and defined by the used fiber as slow- and fast-light medium.

In the following, only the case of slow-light (where g is positive) will be considered for the explanation of the investigations. Nevertheless, the same conditions, but with an reverse behavior, are also valid for the fast-light case (where g is negative). Furthermore, the term *time delay* ΔT is exclusively used for the time delay which is induced by SBS.

The second derivative of wave number k_2 leads to the GVD. Since $k(\omega)$ is complex it can be assumed that the derivatives are complex too. Thus, the imaginary part of k_2 will introduce a further change to the pulse amplitude which is not considered here. The linear term $n\omega/c$ vanishes in the second derivative of $k(\omega)$. Hence, the real part of k_2 is responsible for the GVD which also determines the higher-order dispersions. The frequency-distribution of the GVD is given by:

$$GVD \propto \Re\left(k_2 L\right) = \Re\left[\frac{d^2 k(\omega)\, L}{d\omega^2}\right] = g \frac{\gamma(\omega-\omega_0)^3 - 3\gamma^3(\omega-\omega_0)}{\left[(\omega-\omega_0)^2 + \gamma^2\right]^3}. \quad (3.41)$$

The GVD distribution is rather of qualitative interest for the investigations in this work. The GVD distribution of a single natural Brillouin gain according to Equation (3.41) is shown in Figure 3.7b. In general, the GVD describes the frequency-dependence of the group velocity. As can be seen from Figure 3.7b, the GVD shows a strong variation within the Brillouin gain bandwidth. Hence, every frequency component of the signal pulse will experience another time delay which leads to a change in the pulse shape.

Thus, the SBS-based time delay is accompanied by distortions of the pulse which leads to a broadening of the pulse width. The reasons for this can in fact be found in the GVD and higher-order terms of $k(\omega)$ but especially in the spectral narrowing of the pulse bandwidth due to the narrow Brillouin gain bandwidth. This will be explained in more detail later. The broadening factor for long Gaussian-shaped pulses, as the ratio between the pulse widths of the delayed output pulse to the input pulse, can be described by [183, 184]:

$$B = \frac{\tau_{out}}{\tau_{in}} = \sqrt{1 + g\frac{16\ln 2}{\tau_{in}^2\,\Delta\omega_B^2}} \approx \sqrt{1 + 0.28\frac{g}{\tau_{in}^2\,\Delta f_B^2}} \approx \sqrt{1 + 1.77\frac{\Delta T}{\tau_{in}^2\,\Delta f_B}}. \quad (3.42)$$

3.2 SBS-Based Slow- and Fast-Light

For example, in the experiment of Okawachi et al., the higher fractional delay of the shorter pulse comes at the cost of a higher pulse broadening ($B = 1.05$ for the 63 ns pulse and $B = 1.3$ for the 15 ns pulse) [14].

3.2.2 Principle Experimental Setup

The setup to realize a Brillouin-based slow- and fast-light system, as shown in Figure 3.8, is simple and requires only standard telecommunications components. It consists of three main parts: a system for the optical pulse generation, a pump system to produce the SBS and a measurement system to detect the delayed pulses. These systems are connected by an optical circulator (C) to the slow- and fast-light medium.

The first system generates the optical pulses which are launched into the slow- and fast-light medium, i.e. the fiber under test (FUT), from one side marked as point A. In an existing optical network this would be the interface where the incoming data signals are coupled in. A signal LD, which can be a standard DFB-LD for WDM systems, creates a CW carrier wave at the signal frequency f_s corresponding to a wavelength around 1550 nm. The Gaussian-shaped optical pulses are generated by an external modulation of the carrier wave via a Mach-Zehnder modulator (MZM). To achieve the maximum power at the output of the modulator, the polarization of the pulse at the input has to be adjusted by an optical polarization controller (OPC). The MZM is driven by a pulse generator which supplies the electrical Gaussian-shaped pulses. Throughout the measurements only single pulses with temporal FWHM widths mainly between 30 ns and 42 ns and repetition rates from 0.7 MHz to 1 MHz were used. The operating point of the modulator is adjusted by a bias voltage to convert the pulses from the electrical to the optical domain without impairing their shapes. A variable optical attenuator (VOA) is used to control the amplitude of the signal pulses. To protect all components from backscattered or pump light which propagates in direction of the signal LD, an optical isolator (ISO) is connected to the fiber input A.

From the opposite side of the fiber (point B) the pump waves for the SBS are coupled in via the circulator (port $1 \Rightarrow 2$). The simplest configuration of the pump system only consists of a

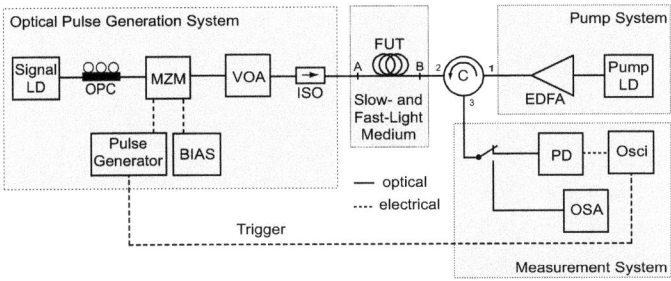

Figure 3.8: Principle experimental setup of Brillouin-based slow- and fast-light systems.

pump DFB-LD and an EDFA to control the pump power. For the optimization of SBS-based slow- and fast-light systems more than one pump wave is used and the pump system is getting more complex. This will be further explained in the next chapters. The pump LD applies a pump wave at the pump frequency f_p. Inside the FUT it creates a Brillouin gain at $f_p - f_B$ and a Brillouin loss at $f_p + f_B$ which lead to an amplification or attenuation and a delay (slow-light) or advancement (fast-light) of the signal pulses. Therefore, the pump and signal waves have to be adjusted so that f_s matches either $f_p - f_B$ or $f_p + f_B$. The time delay is continuously controllable by the EDFA output pump power. For the measurements only SSMF with lengths between 0.1 km and 50 km were used as slow- and fast-light medium.

Finally, the delayed output pulses are detected by the measurement system at the output of the circulator port 3. Here, the SBS process can be monitored by an optical spectrum analyzer (OSA). The time delay and temporal widths of the output pulses are measured by an oscilloscope which is triggered by the pulse generator. For an exact determination of these values the oscilloscope is operated in an average mode of 128 cycles. The optical pulses can either be directly coupled into the optical input of the oscilloscope or first have to be converted from the optical domain back into the electrical domain by a photodiode (PD).

3.2.3 Preliminary Basic Investigations and Measurements

In this section a SBS-based slow-light system with a single natural Brillouin gain is investigated as basis for further considerations [185]. For the investigation SSMF with lengths of 0.1 km, 0.2 km, 0.5 km, 1 km, 2 km, 5 km, 10 km, 15 km, 20 km, 25 km and 50 km were used. All these fibers originate from the same manufacturer, were shortened from one long fiber and were coiled in the same way. Hence, it can be assumed that they have the same properties.

According to Equation (3.40), for the SSMF which show a Brillouin bandwidth of $\Delta f_B = 35$ MHz a time delay slope of approximately 1 ns/dB can be predicted. Considering the correction factor expressed by Equation (3.21) and Equation (3.22), the time delay slope reduces to approximately 0.88 ns/dB. While the time delay as a function of the Brillouin amplifier gain is equal for all analyzed fibers, the time delay as a function of the pump power depends on the fiber length. In general, the time delay decreases if shorter fibers are used. However, according to Equation (3.40), this can be compensated by increasing the pump power. The calculated time delay slopes for the different fiber lengths are listed in Table A.2 and are shown in Figure 3.9a and Figure 3.9b. As can be seen, the slope increases with the fiber length. Therefore, the time delay in [ns] per [mW] pump power increases by approximately $(0.53\,\mathrm{km}^{-1}) \times L_{\mathit{eff}}$. This is due to the smaller Brillouin threshold of longer fibers where the pump depletion and the saturation regime is reached at much lower pump powers. Hence, the time delay is higher in longer fibers if the same pump power is used. This behavior can be seen in Figure 3.9c where the pump power is depicted as a function of the fiber length for several time delay values. For the same time delay the required pump power decreases with the fiber length.

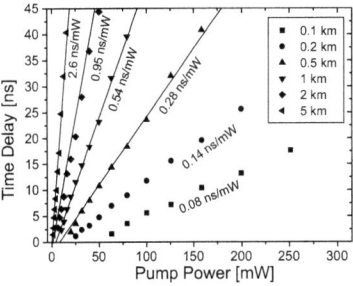

(a) Time delay slope for 0.1 km to 5 km SSMF lengths. The lines show the calculated slopes. The values on the graphs are the measured slopes.

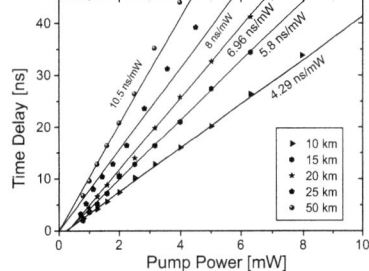

(b) Time delay slope for 10 km to 50 km SSMF lengths. The lines show the calculated slopes. The values on the graphs are the measured slopes.

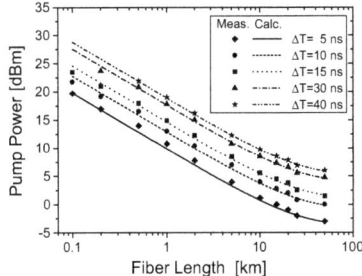

(c) Required pump power as a function of the fiber length for several time delays.

Figure 3.9: Time delay slope and pump power requirements for a SBS-based slow-light system with a single natural Brillouin gain and fibers of different lengths. The parameters for the calculations are: $g_{Bmax} = 2 \times 10^{-11}$ m/W, $K_B = 2$, $A_{eff} = 86\,\mu\text{m}^2$, $\Delta f_B = 35$ MHz and $\alpha = 0.2$ dB/km.

However, this theoretical assumptions are based on a number of simplifications. For example, the time delay in Equation (3.40) is only valid for a small signal gain. Hence, the pump depletion and also any influences of noise were neglected. Furthermore, it is assumed that the fiber properties are uniform over the whole length. Thus, the relation of the polarization states between the pump and signal wave would be maintained theoretically along the SSMF. However, this cannot be applied in practice. Additionally, it is assumed that the Brillouin bandwidth is independent from the pump power and the location in the fiber. However, in reality the Brillouin bandwidth decreases with higher pump powers [99] and depends on the environmental temperature and the mechanical stress which can vary along the fiber length. At last, there exists the claim that the fiber length influences the maximum achievable time delay which can lead to higher obtainable gains and therefore, higher delays in short fibers [186]. These concerns are clarified by a practical verification described next.

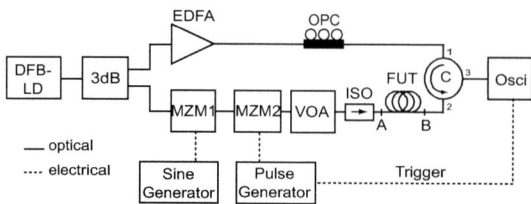

Figure 3.10: Principle experimental setup of a SBS-based slow-light system with a single natural Brillouin gain. The bias voltages and OPC for the MZM are not shown.

The measurement setup for the single natural gain line SBS-based slow-light system is slightly modified from the standard setup explained in the previous section, as can be seen in Figure 3.10. The CW carrier wave with a frequency f_p is used as pump and signal wave simultaneously. Therefore, it is divided into two paths. In the upper or pump path an additional OPC is utilized to equalize the pump and signal polarization states to achieve the highest SBS process efficiency.

In the lower or signal path the carrier wave is externally modulated by MZM1 in suppressed carrier regime with a sinusoidal signal. The modulation frequency is tuned exactly to the Brillouin frequency shift. Hence, the first-order sidebands at $f_p \pm f_B$ are created. Both waves are modulated by a second MZM with the 30 ns long Gaussian-shaped pulses. Inside the FUT the Brillouin gain and loss precisely coincide with the two signal waves, but only the pulses at $f_p - f_B$ are amplified and delayed. The pulses at $f_p + f_B$ vanish due to the Brillouin loss. Since the pump and the signal wave were generated by the same laser source, they are perfectly matched with a high frequency stability. For the sake of clarity the used OPC and bias voltages for the modulators are not shown in the setup.

Next to the time delay and the pulse width, the Brillouin amplifier gain was also measured. Therefore, the input pulse power was held constant at −60 dBm throughout all measurements. The gain factor, disregarding the fiber attenuation, was determined by the optical power ratio of the output pulses with and without SBS.

The measured time delay as a function of the Brillouin gain is shown in Figure 3.11a. As can be seen, by increasing the gain the time delay increases linearly as well. In general the diagram shows that the time delay is maintained in fibers with different lengths as long as the same amount of gain is provided. An increase in the SBS gain and hence, a larger time delay in shorter fibers was not observed. Thus, the time delay increases similarly by 0.8 ns per dB gain for all measured fibers and large fractional time delays ΔT_{frac} of up to approximately two can be achieved. The predicted slope of 0.88 ns/dB was not perfectly matched. The reasons for the slight deviation may be attributed to random polarization fluctuations and the gain-dependence of the Brillouin bandwidth along the SSMF [99]. The measured slope values of the time delay as a function of the pump power are listed in Table A.2 and are attached to the graphs in

3.2 SBS-Based Slow- and Fast-Light

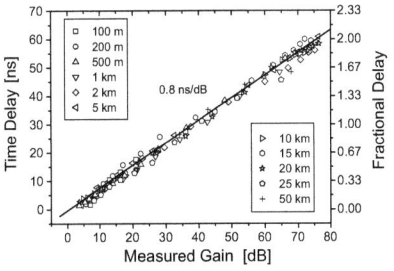 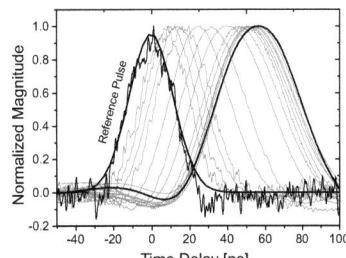

(a) Time delay as a function of the Brillouin gain factor for the different fiber lengths.

(b) Normalized time functions of delayed pulses and reference pulse in a SSMF with a length of 5 km.

Figure 3.11: Measured time delay and time functions of the delayed pulses.

Figure 3.9a and Figure 3.9b. As can be seen, the predicted and the measured slopes are in a good agreement. Hence, the background theory described in Subsection 3.2.1 has been verified by the experiments. Since only one pump wave is present and the gain threshold as well as the delay-gain slope is equal for all fiber lengths, the maximum induced gain and time delay depend on the power of the input pulse [164]. The fiber length is not crucial for the SBS delay time. Thus, maximum achievable time delays of approximately 9 ns and 65 ns can be estimated for input pulse powers of 0 dBm and −60 dBm for instance [164].

In Figure 3.11b the normalized time functions of the delayed pulses in comparison to the reference pulse are additionally shown for a SSMF with a length of 5 km. The pulse diagrams of the other fibers can be found in Figure B.1 of Section B.1. The reference pulse is defined as the output pulse without the SBS effect and was Gaussian fitted for an exact determination of the time delays and pulse widths. The pulse with the maximum time delay is shown as black bold line on the right hand side of the diagram. If this pulse is compared with the reference pulse it can be clearly seen that the time delay is accompanied by a broadening of the pulse width.

The broadening and the output pulse widths as a function of the pump power are shown in Figure 3.12. As can be seen in the diagram, the measured and calculated values deviate partially from each other. One reason for this is that the theoretical prediction of the pulse broadening according Equation (3.42) is only based on the bandpass filtering caused by the limited bandwidth of the SBS and that third and higher-oder dispersion are neglected [4, p. 40]. Furthermore, the averaging of the pulses during the measurement process leads to minor corruptions of the pulse width values. However, Figure 3.12 shows in general that the power-dependent broadening factor is different for different fibers lengths as for the time delay. In longer fibers, where the time delay increases rapidly with the pump power, the output pulse width also rises very quickly. Therefore, broadening factors of on average 1.5 relating to input pulse widths of 30 ns can occur in such a standard slow-light system. Due to this value of B the achievable effective time delay ΔT_{eff} is reduced to approximately 67 % for a fractional delay

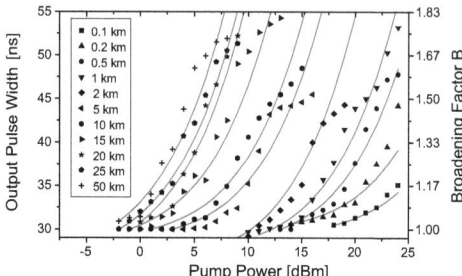

Figure 3.12: Pulse broadening as a function of the pump power in a SBS-based slow-light system with a single natural Brillouin gain and variable SSMF length. The lines show the calculated broadening and the symbols represent the measured values. The parameters for the calculations are: $g_{Bmax} = 2 \times 10^{-11}$ m/W, $K_B = 2$, $A_{\mathit{eff}} = 86\,\mu\mathrm{m}^2$, $\Delta f_B = 35$ MHz, $\alpha = 0.2$ dB/km and $\tau_{in} = 30$ ns.

of one. Hence, the pulse broadening is one crucial limiting parameter in SBS-based slow-light systems.

3.3 Properties of SBS-Based Slow- and Fast-Light

In this section the most important properties of standard SBS-based slow- and fast-light systems are summarized. In general, the SBS has several significant advantages compared to other slow- and fast-light methods:

1. For the time delay only small pump powers of a few milliwatts are required. Since the time delay depends directly on the applied pump power, it can be continuously tuned over a wide range. Moreover, the SBS offers the opportunity to achieve high fractional time delays up to twice of the initial pulse width, as shown in Figure 3.11a.

2. The SBS works at room temperatures and in all fiber types within their whole transparency range. It operates at all wavelengths, especially at those which are used in optical communications. Thereby, the input and output wavelengths of the signal pulses are maintained. No wavelength conversion is applied. Furthermore, the fiber is used as Brillouin and slow-light medium simultaneously. Thus, the incoming fiber in an optical network can directly be used for the delay of the signals.

3. For the SBS-based slow- and fast-light setups reliable off-the-shelf telecommunications components can be used. This keeps the systems simple, cheap, robust and easy to implement into existing networks.

Besides these advantages, the SBS also suffers from some limitation:

1. The narrow Brillouin bandwidth of around 35 MHz limits the maximum delayable distortion-free data rates to a few tens of Mbit/s which are not contemporary for today's data

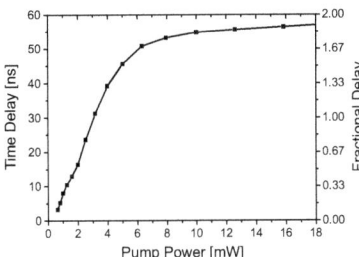

Figure 3.13: Saturation of the time delay in a SSMF with a length of 25 km.

transmissions. Conventional system bandwidths should provide data rates in the Gbit/s-range and beyond.

2. The time delay of the pulses is accompanied by a strong broadening of the pulse width which reduces the effective time delay, as shown in Figure 3.12. This is mainly induced by the narrow Brillouin bandwidth which acts like a bandpass filter and leads to a spectral narrowing of the pulse spectrum. Furthermore, the GVD and higher-order terms of the wave number have an influence on the output pulse width.

3. The maximum achievable time delay saturates at high pump powers and hence, is limited by the pump depletion. In the saturation regime the Brillouin amplifier gain and the time delay does not increase any further, as can be seen in Figure 3.13.

To resolve these drawbacks and to improve the SBS-based slow- and fast-light performance the research and this work have concentrated on three particular priorities: enable broadband SBS, time delay enhancement and distortion reduction. The basis of such an optimization of SBS-based slow- and fast-light systems is the simple tailoring of the Brillouin spectrum by designing the pump spectrum. This unique feature will be explained in detail in the following chapters.

3.4 Summary

The fundamentals and first investigations on slow- and fast-light based on SBS were described in this chapter. At first the nonlinear effect of SBS was described in detail as it is the key method to produce the pulse delay or advancement used in this work. Therefore, the generation and mode of operation were explained in general. Furthermore, important properties and parameters, such as threshold, gain, spectral shape and bandwidth, were defined and determined experimentally.

After that, the creation of slow- and fast-light due to Brillouin scattering was explained in theory and verified experimentally. It was shown that slow-light can be achieved within a Brillouin gain resonance whereas fast-light is related to a Brillouin loss. In both cases the alteration of the group velocity is scaled with the pump power and hence, with the available gain or loss. In a preliminary measurement with a single natural Brillouin gain in SSMF with lengths between 100 m and 50 km it was shown that the time delay generally increases with the pump power. Therefore, the same time delay can be achieved in every fiber length as long as the same gain is available.

At the end of this chapter the properties of SBS-based slow- and fast-light systems were summarized. On the one hand, these systems have many significant advantages. On the other hand, the systems suffer from some serious disadvantages which limit the slow-light performance. Thus, one main objective of the work is to resolve these challenges by investigating different methods for Brillouin bandwidth enhancement, time delay enhancement and pulse distortion reduction. These methods are based on the unique feature which is that the Brillouin spectrum can be tailored by a modulation of the pump spectrum.

4 Brillouin Bandwidth Enhancement

One line of research on SBS-based slow- and fast-light has focused on broadband SBS where the limit of the natural bandwidth is circumvented. In this chapter the fundamentals of a Brillouin bandwidth broadening are described and two methods to achieve a broadband SBS spectrum are discussed. This includes also experimental results which show that the slow- and fast-light bandwidth can be suitable for high bit rate data transmission.

4.1 Introduction

As explained in the previous chapter, the narrow Brillouin bandwidth is one limiting factor of SBS-based slow- and fast-light systems. For a monochromatic pump wave typical bandwidths from only 10 MHz to 50 MHz are obtainable in SSMF at wavelengths around 1550 nm [158]. On the one hand, this restricts the maximum delayable data rate and on the other hand this leads to a spectral narrowing of the pulse spectrum and therefore, to a pulse distortion. However, as described in Subsection 3.1.4, the effective Brillouin spectrum $g_B^{\mathit{eff}}(f)$ is due to the convolution of the pump spectrum $P_p(f)$ and the natural Brillouin spectrum $g_B(f)$ [162, 187]:

$$g_B^{\mathit{eff}}(f) = g_B(f) \otimes P_p(f). \tag{4.1}$$

The convolution of the natural Brillouin spectrum and a possible pump spectrum is shown in Figure 4.1. As can be seen, the convolution results in a Brillouin spectrum which merges the shape of the polychromatic pump spectrum. Therefore, every peak of the discrete pump spectrum creates its own intrinsic Brillouin gain which are overlapping in the fiber. Hence, this fundamental approach can be used to achieve an enhancement of the Brillouin bandwidth. However, as derived from Equation (3.40), the broadening decreases the gain and hence, the absolute time delay in the same order of magnitude [17]. At the same time the pulse width broadening is decreased, as can be seen from Equation (3.42). Therefore, the effective time delay in the non-saturated regime is maintained. However, higher pump powers are necessary. Furthermore, the time delay decrease due to the Brillouin bandwidth enhancement cannot be compensated completely [188], as will be shown in Section 6.2.

Basically, there are two opportunities to enhance the Brillouin bandwidth:

1. using a polychromatic pump with discrete spectral lines with a certain distance [189–192] and

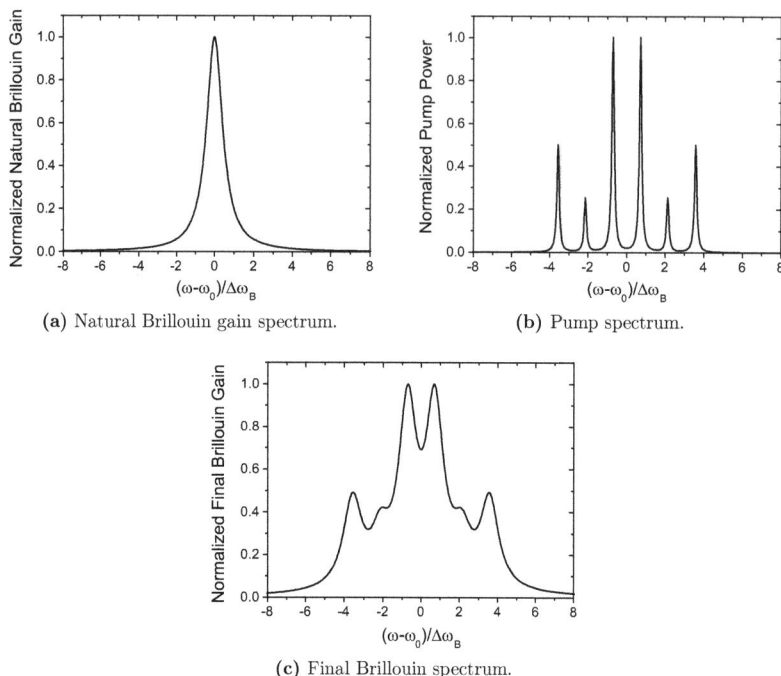

Figure 4.1: Convolution of the pump spectrum with the natural Brillouin gain spectrum.

2. using a (single or multiple line) pump with a linewidth broader than the intrinsic Brillouin bandwidth [97, 187, 193–195].

In the first method, every spectral pump line produces a separate Brillouin gain in the fiber. Hence, the gains will overlap to the final broadened Brillouin spectrum if the frequency distance of the pump waves lies in the range of the intrinsic Brillouin bandwidth [189]. The different pump lines can be created by using independent monochromatic pump lasers for instance. Another opportunity is to apply an external modulation of a single pump LD. For example, two independent LD were externally phase modulated by a sinusoidal signal with a frequency of 28 MHz to broaden the natural Brillouin bandwidth of 28 MHz by a combination of six independent Brillouin gains [189, 190]. Therefore, the power of the carriers and modulation sidebands was adjusted to be equal. As a result the Brillouin spectrum was enhanced by a factor of approximately 5 to an overall uniform spectrum with a bandwidth of 144 MHz.

Recently, slow-light with a 200 MHz broadband flat-top Brillouin spectrum has been demonstrated using a gain pump light consisting of 20 discrete line spectra [191]. Up to now, the highest bandwidth obtained with this approach is approximately 300 MHz [192]. Although the Brillouin bandwidth is broadened by a factor of 10, this has remained insufficient for Gbit/s data transmissions.

Instead of an analog modulation with a sinusoidal signal also digital modulation formats are possible to enhance the Brillouin bandwidth. For example, a binary phase-shift-keying (BPSK) was used to obtain a Brillouin FWHM bandwidth of 1.5 GHz [140]. Therefore, arbitrary bit patterns with a defined transition probability between the inverse logical states of two adjacent bits have been applied as modulation signal.

For the second method a pump source with a broadband emission spectrum is necessary. This can be facilitated by a direct modulation of the injection current of a single LD for instance. In doing so, a Brillouin bandwidth of 150 MHz was achieved [187]. Later on, the 50 MHz intrinsic Brillouin bandwidth of a dispersion-shifted fiber (DSF) has been broadened up to approximately 325 MHz by modulating the pump source's injection current with a pseudo-random binary sequence (PRBS) [193]. Within this gain bandwidth 2.7 ns wide pulses have been delayed up to approximately 2.75 ns. This work was further improved by modulating the pump source with a Gaussian white noise. Therefore, a SBS slow-light bandwidth as large as 1.9 GHz was achieved, thereby supporting a data rate of nearly 1 Gbit/s and delaying pulses as short as 400 ps up to 1 ns [194]. The highest reported SBS slow-light bandwidth which was achieved by using a single broadband modulated pump source is as large as 12.6 GHz [195]. Thereby, data rates of over 10 Gbit/s are supported and 47 ps of time delay for 75 ps long pulses was shown.

Besides a modulated pump, also other sources with a natural broadband emission spectrum can be used for a Brillouin bandwidth enhancement. For instance ASE sources can provide bandwidths up to tens of GHz for that purpose. For example, a continuous delay of two independent 2.5 Gbit/s signals by as much as 170 ps at a BER of 10^{-9} has been shown within a single slow-light medium which was pumped by incoherent spectrally sliced ASE sources with bandwidths of up to 4 GHz [97].

4.2 Experimental Verification and Results

In this section both opportunities for the Brillouin bandwidth enhancement are verified experimentally. Thereby, the use of a broadband pump is more efficient to achieve high bandwidths than the use of a polychromatic pump source. However, besides the fundamental demand for higher bandwidths both methods are of special practical importance because they are also capable for the reduction of the delay-induced pulse distortions, as will be shown in Chapter 6.

4.2.1 Polychromatic Pump Sources

The experiment to achieve the first scheme of a broadened reshaping of the Brillouin spectrum by using a polychromatic pump wave consisting of a discrete line spectrum was accomplished according to [189, 190]. The experimental pump setup is shown in Figure 4.2. The pump lines are produced by an external phase modulation of a single LD. Therefore, the phase modulator (PM) is driven by a sinusoidal signal. The input polarization of the modulator is adjusted

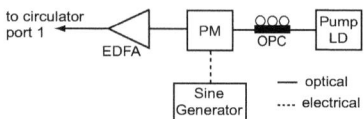

Figure 4.2: Experimental pump system for Brillouin bandwidth enhancement via external modulation of the pump source.

by an OPC to achieve the maximum output power behind the PM. Due to the modulation a frequency comb is produced whose optical power is controlled by an EDFA. By adjusting the electrical power of the modulation signal the frequency comb can be flattened. With a modulation frequency in the range of the intrinsic Brillouin bandwidth a broadened flat-top uniform Brillouin spectrum is achieved. The bandwidth of the final spectrum depends on the number of frequency components in the comb. By using multiple external modulated pump sources this method can be further improved and adapted to any given application [189, 190].

This approach was used to reduce the distortions in a cascaded SBS-based slow-light system [196], as will be explained in Chapter 6. In Figure 4.3a the overlapping of six independent Brillouin gains is illustrated. For the simulation a natural Brillouin bandwidth of 24 MHz was assumed. The overlapping leads to a broad and flat-top Brillouin gain which is translated into a broadened distribution of the group index change on the one hand. On the other hand, as can be seen from the characteristic of the refractive index, the SBS induced dispersion is decreased within the overall Brillouin bandwidth. Thus, the broadening leads to a reduced time delay as well. The broadened spectrum was measured using the methods described in Section A.3. In Figure 4.3b the final Brillouin spectrum in comparison to the natural spectrum can be seen. Therefore, the FWHM bandwidth was broadened to approximately 144 MHz.

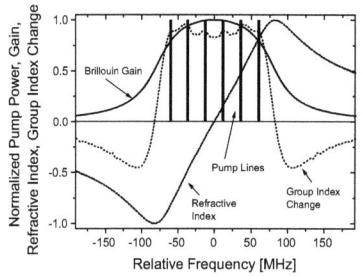

(a) Gain, refractive index and group index change.

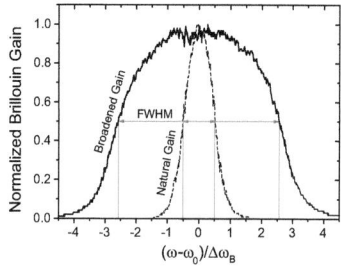

(b) Measured broadened Brillouin gain. (From [154, 190] with permission.)

Figure 4.3: Brillouin bandwidth enhancement via multiple discrete pump lines.

4.2.2 Broadband Pump Sources

The experimental pump setup to achieve a Brillouin bandwidth enhancement using a broadband pump source is shown in Figure 4.4. The pump DFB-LD was directly modulated with a Gaussian noise signal to generate a broad Brillouin bandwidth with a Gaussian shape. Since, the broadening of the pump spectrum reduces the Brillouin gain, the pump power has to be amplified by an EDFA. Due to the direct modulated pump spectrum the SBS spectrum is continuously broadened, as can be seen in Figure 4.5a. Hence, with this bandwidth enhancement method it is more efficient and much simpler to achieve an ultra-wideband SBS spectrum than when using a polychromatic pump source with discrete spectral lines. The measured spectrum of the broadened gain with a FWHM bandwidth of approximately 10 GHz is shown in Figure 4.5b. It was achieved by a direct modulation of the DFB-LD with a noise with a power of 30 dBm [154].

The maximum Brillouin bandwidth, which can be achieved by this method, is reached if the broadening values are in the region of the Brillouin shift, i.e. around 10 GHz. As can be seen from Figure 4.5a and Figure 4.5b the broadening of the Brillouin gain around $f_p - f_B$ also leads to a broadening of the Brillouin loss around $f_p + f_B$. Hence, a further enlargement of the pump spectrum leads to a compensation between the loss and the gain in the region of f_p. Thereby, the maximum achievable Brillouin bandwidth depends on the shape of the SBS spectrum. Thus, the bandwidth can take on values which correspond to f_B (for Gaussian-shaped Brillouin

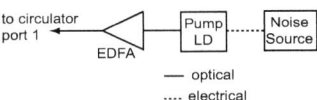

Figure 4.4: Experimental pump system for Brillouin bandwidth enhancement via direct modulation of the pump source.

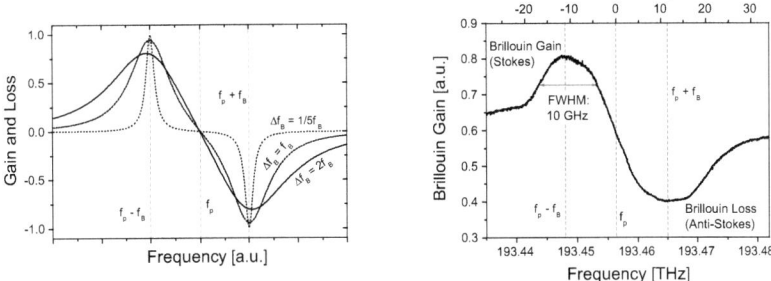

(a) Brillouin gain and loss spectra for different broadening values. (From [197] with permission.)

(b) Measured broadened Brillouin gain. (From [154] with permission.)

Figure 4.5: Brillouin bandwidth enhancement via direct modulation of one pump source.

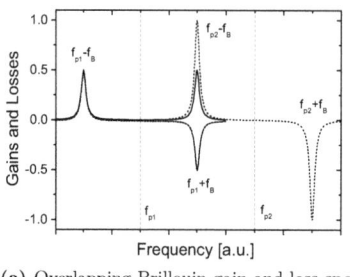
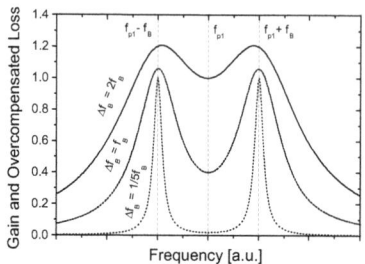

(a) Overlapping Brillouin gain and loss spectra.

(b) Final Brillouin gain spectrum for different broadening values.

Figure 4.6: Idea of Brillouin bandwidth enhancement via direct modulation of two pump sources. (From [197] with permission.)

spectra for instance) up to $2f_B$ (for ideal rectangular-shaped Brillouin spectra) [195, 197, 198]. According to this, with a SBS-based slow-light delay system data rates of around 10 Gbit/s can be delayed at most. However, in optical communications bit rates of 40 Gbit/s and more are commonly used [197]. Thus, a further Brillouin bandwidth enhancement is required.

However, it has been shown that this limitation is only valid if a single broadband pump source is used and can be circumvented by using a second pump at a frequency $f_{p2} = f_{p1} + 2f_B$ for instance [197]. Such a configuration is illustrated in Figure 4.6a. As can be seen, the Brillouin loss of the first pump source coincides with the Brillouin gain of the second pump source at $f_{p1} + f_B = f_{p2} - f_B$. On the one hand, if both spectra have the same magnitudes and bandwidths the loss and the gain cancel one another. On the other hand, if the pump power of the second laser is higher than that of the first laser its loss can be overcompensated. Due to the fact that the overall gain is the superposition of the individual gains a very broad slow-light bandwidth will be generated if the spectrum of each pump laser is broadened up to the Brillouin shift and higher [197]. Therefore, the second gain should be twice of that of the first to achieve a flat-top and uniform overall gain distribution. The simulated broad overall Brillouin spectrum, consisting of the overlapping of a single broadened gain and the overcompensated loss, is shown in Figure 4.6b.

A further enhancement of the bandwidth up to a few tens of GHz can be expected if more than two pump lasers are incorporated in the same manner [197]. However, the overcompensation of the losses needs increasing pump powers which drive the system into saturation.

The experimental setup for the pump system of this concept is shown in Figure 4.7c. Two separate pump LD are directly modulated by a noise source. The noise signal is divided into two equal parts to provide the same broadening of both pumps. A tunable optical coupler controls the relation of the optical output power of these two pump lasers. Thus, the pump power of LD2 can be chosen to be twice the pump power of LD1 to enable the overcompensation of the Brillouin loss. An EDFA amplifies and controls both pump waves while their power relation

4.2 Experimental Verification and Results

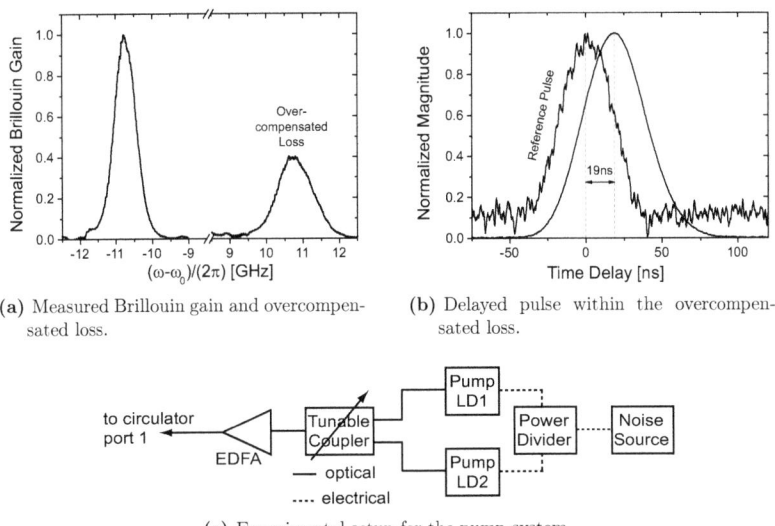

Figure 4.7: Practical Brillouin bandwidth enhancement via direct modulation of two pump sources. (From and after [197] with permission.)

remains the same. Since the broadening and the overcompensation requires a high pump power, the used EDFA was driven in the region of its maximum output power of 23 dBm. After this, the pump waves are launched into the slow-light medium via the circulator. To provide the overcompensation of the loss due to LD1 with the gain of LD2, the distance of the carrier frequencies of the two lasers is adjusted exactly to twice the Brillouin shift.

The measured overall Brillouin spectrum around f_{p1} is shown in Figure 4.7a. As can be seen, the loss spectrum of the first pump laser is in fact overcompensated by the second one. On both sides of LD1 broad positive gains with bandwidths of approximately 2 GHz occur. Due to different direct modulation properties of the laser mounts the SBS bandwidths and magnitudes show different values [197]. Unfortunately, the available experimental equipment was insufficient to broaden the Brillouin gain further. On the one hand, the available pump and noise power were too small to achieve broader SBS gain bandwidths. On the other hand, the tuning sensitivity of the LD limits the maximum noise bandwidth and power which can be supplied to the lasers. Therefore, a further broadening would have lead to damages on the equipment.

To prove that the overcompensated loss also shows a slow-light behavior, pulses with a temporal FWHM of 40 ns were delayed at $f_{p1} + f_B$. For this, the noise source was switched off because such a pulse does not require a bandwidth exceeding the natural Brillouin bandwidth of approximately 35 MHz. Hence, the overcompensated loss has a spectral width which corresponds to the intrinsic Brillouin linewidth. One example of a delayed pulse in comparison to the reference pulse is shown in Figure 4.7b. For a pump power of 12.7 dBm the pulse was delayed by 19 ns inside the overcompensated loss.

The same idea of overlapping the Brillouin spectra of two pump sources and overcompensating the loss of the one pump with the gain of the other pump, but in two different fibers (a 2 km high-NA fiber and a 6 km DSF, both with $f_B \approx 10.5$ GHz), was realized later by Song and Hotate [199]. An enhanced Brillouin bandwidth of up to 25 GHz was achieved and a variable time delay of up to 10.9 ps with 37 ps pulses within this bandwidth was reported.

4.3 Summary

The first line of research was the enhancement of the narrow Brillouin bandwidth to enable ultra-wideband slow- and fast-light. Therefore, two methods were investigated in this chapter. In general, both techniques are based on the nature of SBS that the final Brillouin spectrum results from the convolution between the pump spectrum and the intrinsic Brillouin spectrum. Thus, by broadening the pump spectrum the Brillouin bandwidth can be significantly enhanced which makes SBS-based slow- and fast-light systems compatible with high data rate signals.

The first method is based on the use of a polychromatic pump source which consists of discrete spectral lines with a certain distance. In the slow-light medium every pump line creates an independent Brillouin gain which overlap to an overall Brillouin gain whose bandwidth is broadened. With this method a Brillouin FWHM bandwidth of 144 MHz was achieved. However, since the method is not restricted to one pump source, a further enhancement is possible by using multiple external modulated pump sources.

The second method, which is much more effective, is based on the use of a broadband pump source with a linewidth broader than the intrinsic Brillouin bandwidth. Such a pump source was realized by a direct modulation of the pump laser with a noise signal. If only one pump source was used a maximum Brillouin FWHM bandwidth of approximately 10 GHz was achieved. A further enhancement is restricted because the simultaneously broadened Brillouin loss cancel with the broadened Brillouin gain. This constraint can be mitigated if a second directly modulated pump source is introduced. This was adjusted in a manner that its gain overcompensates the loss and overlaps with the gain of the first pump source. Hence, the Brillouin bandwidth can be further enhanced to approximately twice the Brillouin shift.

These results showed that SBS-based slow-light systems are applicable with data rates of tens of Gbit/s. However, the fractional time delays < 1 are still too small in such ultra-wide bandwidth systems. Hence, reliable methods to improve and to enhance the time delays in SBS-based slow-light are necessary. This will be described and discussed in detail in the next chapter.

5 Time Delay Enhancement

In this chapter the second main field of work about the time delay enhancement in SBS-based slow-light systems is described. After a short introduction three investigated methods for the improvement of the time delay are explained and discussed: the superposition of a narrow Brillouin gain with a broadband loss, the superposition of a Brillouin gain with two narrow losses at its spectral boundaries and the suppression of spurious backscattered Stokes-waves by using short fibers and optical filters in multiple-pump-line systems. The following sections include the theoretical and mathematical background as well as the experimental realization and results of those methods.

5.1 Introduction

A broadened Brillouin spectrum, as described in the previous chapter, shows that SBS-based slow- and fast-light can be applied to data rates in the Gbit/s range. Nevertheless, the increase in SBS bandwidth reduces the gain and the time delay, as shown by Equation (3.40). On the one hand, this can be compensated by higher pump powers. On the other hand, since the delay is combined with an inherent amplification of the pulses the SBS saturates and a further increase of the time delay is restricted if the pump power exceeds the Brillouin threshold, as shown in Figure 3.13. According to Equation (3.16) the gain threshold is approximately 82 dB. Hence, maximum time delays of up to 65 ns could only be achieved with a single gain line SBS-based slow-light system. For example, in the first experiments of SBS-based slow- and fast-light maximum time delays of only between 25 ns and 30 ns have been achieved [13, 14]. Therefore, the corresponding maximum fractional time delays have been only between 0.3 and 1.3 which is to low for an intermediate storage of signals in optical communications. Thus, for practical applications this limitation has to be overcome.

The easiest way to enhance the maximum achievable time delay is to cascade several standard SBS-based delay lines, such as the system shown in Figure 3.8. In doing so, the time delay of every stage is added up. Therefore, the signal power between every stage has to be attenuated to produce the maximum Brillouin amplification and hence, the maximum time delay. This idea was proposed by Song et al. [184]. They cascaded four fiber segments and achieved a large continuous delay of more than 150 ns. However, besides the time delay also the broadening of the delayed pulse is added up in every stage simultaneously. Thus, the pulses were broadened from their initial temporal width of 42 ns to a considerable 102 ns. Therefore, the large fractional delay of 3.6 was strongly decreased to an effective time delay of 1.5. Although the pulse broadening can be reduced [196], as it is described in Chapter 6, this approach has become less promising for the time delay enhancement because it has further disadvantages. The systems are very complex as every cascade is pumped separately by its own source. This needs a significant amount of equipment because additional amplifiers, attenuators, couplers and circulators become necessary. Furthermore, each segment has to be adjusted separately if the SBS properties of the fibers, especially the Brillouin shift, are different in every stage. This leads to a considerably increased effort.

A much simpler idea to improve the time delay efficiency of SBS-based slow-light systems is to decouple the time delay from the Brillouin amplification process and therefore, from the pump depletion. The Brillouin amplification and the saturation depends on the maximum value of the Brillouin gain distribution [200, 201] which in turn is defined by the Brillouin threshold and the peak value of the Brillouin gain coefficient g_{Bmax}, respectively. Hence, if the maximum value of the gain distribution will be decreased the saturation is reduced and higher applicable pump powers are enabled. To achieve this feature, a narrow Brillouin gain can be superimposed with a broad Brillouin loss. The result of this is a transparency window within an absorption band which is in principle similar to the concept of EIT, as described in Subsection 2.4.2.

Independently from the investigations followed in Section 5.2 also another research group developed the same technique [202]. In contrast to the time delay enhancement this was primarily used to produce zero-gain slow- and fast-light. For this, the Brillouin gain was shifted by loss so that the absolute gain value is zero. Thus, the delayed pulses are not amplified and their input amplitude remains constant.

While the saturation depends on the maximum magnitude of the gain, the time delay is a function of the gradient of the Brillouin gain distribution [200, 201]. If the gradient is increased the time delay can be enhanced. Such an enhancement of the delay at the center of a Stokes resonance arises for example from the spectral boundary of an anti-Stokes resonance if it is introduced at the spectral boundary of the Stokes gain [195]. Moreover, if the Brillouin gain is superimposed with two losses at both of its spectral boundaries the delay can be increased significantly [165]. This is another method which can be used for a time delay improvement and enables higher delays in comparison to the superposition of a narrow Brillouin gain with a broadened Brillouin loss.

Hence, the tailoring of the Brillouin spectrum via superpositions of several Brillouin gain and loss spectra can be used for a drastic enhancement of the time delay. The investigation of these opportunities is explained in the following sections.

5.2 Superposition of a Narrow Brillouin Gain With a Broadband Brillouin Loss

The first opportunity to enhance the delay is to decouple the time delay from the amplification gain by superimposing a narrow Brillouin gain with a broad loss. Therefore, the theoretical background and experimental results are described in the following subsections.

5.2.1 Background Theory

The complex wave number for a superposition of a narrow Brillouin gain g_1 with the angular HWHM bandwidth γ_1 and a broad Brillouin loss spectrum $-g_2$ with the angular HWHM bandwidth γ_2 can be written as [165]:

$$\begin{aligned} k(\omega) &= \frac{n}{c}\omega + \frac{1}{2L}\left[\frac{g_1\gamma_1}{(\omega-\omega_0)+j\gamma_1} - \frac{g_2\gamma_2}{(\omega-\omega_0)+j\gamma_2}\right] \\ &= \frac{n}{c}\omega + \frac{1}{2L}\left\{\frac{g_1\gamma_1(\omega-\omega_0)}{(\omega-\omega_0)^2+\gamma_1^2} - \frac{g_2\gamma_2(\omega-\omega_0)}{(\omega-\omega_0)^2+\gamma_2^2}\right. \\ &\quad \left. -j\left[\frac{g_1\gamma_1^2}{(\omega-\omega_0)^2+\gamma_1^2} - \frac{g_2\gamma_2^2}{(\omega-\omega_0)^2+\gamma_2^2}\right]\right\}, \end{aligned} \quad (5.1)$$

where $g_{1,2} = g_B P_{p1,2} L_{\mathit{eff}}/A_{\mathit{eff}}$ with $P_{p1,2}$ as pump powers for the gain and the loss. Hence, the complex transfer function of the SBS-based slow-light system becomes:

$$\begin{aligned} H(\omega) &= \exp\left\{\frac{g_1}{2}\frac{\gamma_1^2}{(\omega-\omega_0)^2+\gamma_1^2} - \frac{g_2}{2}\frac{\gamma_2^2}{(\omega-\omega_0)^2+\gamma_2^2}\right. \\ &\quad \left. +j\left[\frac{g_1}{2}\frac{\gamma_1(\omega-\omega_0)}{(\omega-\omega_0)^2+\gamma_1^2} - \frac{g_2}{2}\frac{\gamma_2(\omega-\omega_0)}{(\omega-\omega_0)^2+\gamma_2^2} + \frac{n}{c}\omega L\right]\right\}. \end{aligned} \quad (5.2)$$

The real part of the exponent in Equation (5.2) describes the superposition of the narrow Lorentzian-shaped Brillouin gain with a broad Lorentzian-shaped Brillouin loss. Normally, the broadened loss would show a Gaussian distribution as described in Subsection 3.1.4. However, since the loss spectrum is only used for the delay-gain-decoupling and its shape has hardly influence on the time delay, this is neglected here to simplify the discussion. From $2\Re[jk(\omega)L]$ the frequency-distribution of the overall gain can be derived which is shown in Figure 5.1a. As can be seen, the base line of the natural SBS gain is shifted into a loss region. Thus, the maximum of the overall gain in the line center is also reduced by $g_1 - g_2$. In contrast to this, the shape of the gain is almost maintained within the FWHM bandwidth. For the simulation the loss bandwidth is five times higher than the gain bandwidth. If the loss is much broader than the gain the Brillouin gain spectrum will only be shifted into the loss region whereas its

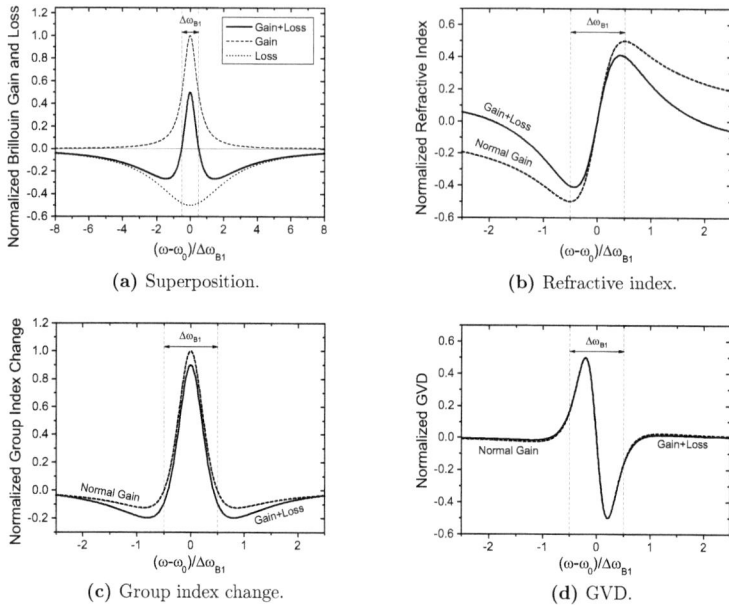

Figure 5.1: Behavior of a superposition (solid) of a gain (dashed) with a broadened loss (dotted) in comparison to a natural Brillouin gain with $g_1 = 2g_2$ and $\Delta\omega_{B2} = 5\Delta\omega_{B1}$. All parameters are normalized to the natural Brillouin gain case.

shape remains almost the same. Since the group index and the group velocity depends on the spectral shape, the time delay will also be unchanged. At the same time, the pump depletion is reduced due to the lower Brillouin gain. This enables the opportunity to use higher pump powers which in turn leads to higher achievable time delays when compared to a single Brillouin gain.

The imaginary part of the exponent in Equation (5.2) leads to a phase change and results in a change of the refractive index. The frequency-distribution of the refractive index of the superimposed gain in comparison to a normal gain is shown in Figure 5.1b. Both graphs show a positive dispersion with an almost equal slope within the FWHM bandwidth. Hence, the group index change, which is the derivative $d\{\Im[jk(\omega)L] - n\omega L/c\}/d\omega$, shows almost the same characteristic for both cases as well, as can be seen in Figure 5.1c. Thereby, the curve for the superimposed gain is only slightly lower although the absolute gain is strongly reduced. However, the higher the relation between the loss and the gain bandwidth, the better the group index distribution of the superimposed gain approximates to the group index distribution of the normal gain. Hence, for the same time delay a much lower amplification gain is necessary and the time delay can be enhanced due to higher usable pump powers.

5.2 Superposition of a Narrow Brillouin Gain With a Broadband Brillouin Loss

The first derivative of $k(\omega)$ yields:

$$\frac{dk(\omega)}{d\omega} = k_1 = \frac{1}{v_g} = \frac{n}{c} - \frac{1}{2L}\left\{\frac{g_1\gamma_1}{[(\omega-\omega_0)+j\gamma_1]^2} - \frac{g_2\gamma_2}{[(\omega-\omega_0)+j\gamma_2]^2}\right\}, \quad (5.3)$$

with

$$\Re(k_1) = \frac{n}{c} + \frac{1}{2L}\left\{g_1\frac{\gamma_1^3 - \gamma_1(\omega-\omega_0)^2}{\left[(\omega-\omega_0)^2 + \gamma_1^2\right]^2} - g_2\frac{\gamma_2^3 - \gamma_2(\omega-\omega_0)^2}{\left[(\omega-\omega_0)^2 + \gamma_2^2\right]^2}\right\}. \quad (5.4)$$

Hence, the SBS time delay in the line center of the Brillouin spectrum ($\omega = \omega_0$) becomes:

$$\Delta T = \frac{g_1}{2\gamma_1} - \frac{g_2}{2\gamma_2} = \frac{g_1}{2\pi\,\Delta f_{B1}} - \frac{g_2}{2\pi\,\Delta f_{B2}}, \quad (5.5)$$

and with $k = \gamma_2/\gamma_1$ as the ratio between the loss and gain bandwidths:

$$\Delta T = \frac{1}{2\gamma_1}\left(g_1 - \frac{g_2}{k}\right) = \frac{1}{2\pi\,\Delta f_{B1}}\left(g_1 - \frac{g_2}{k}\right). \quad (5.6)$$

As can be seen from Equation (5.5) and Equation (5.6), the time delay is generally reduced by the additional loss and increases slower with the amplifier gain in comparison to the normal gain case. However, since the delay is decoupled from the amplifier gain at the same time, higher gain pump powers can be used which lead to an enhancement of the time delay. If for example $g_1 = g_2$, the final Brillouin amplifier gain in the line center is $g = g_1 - g_2 = 0$. Thus, the delay process will not change the amplitude of the delayed pulses. If in addition the gain and loss have equal bandwidths, $k = 1$ and the time delay becomes zero because the loss and gain cancel each other. If $0 < k < 1$ the time delay becomes negative which leads to fast-light. This is further explained in Chapter 7. If the loss bandwidth is much broader than the gain bandwidth ($k \gg 1$), according to Equation (5.6), the time delay slope will approximate to the normal gain case. Hence, by this method the same time delays as with a single natural Brillouin gain can be achieved, but without a change of the amplitude of the output pulses [202]. In practice, much higher pump powers are required for this. However, the maximum possible amplifier gain $g_{th} = 19$, which is responsible for the amplification of the pulse amplitude and which is defined by the Brillouin threshold, remains unchanged. Therefore, the absolute overall gain can be as high as the normal gain and the time delay becomes higher if the loss is overcompensated [164]. For $g_1 = g_2$ this would result in a relative gain of $2 \times 19 = 38$, which is responsible for the time delay.

The great advantage of this method is that the frequency-distribution of k_1 and hence, the group index change is almost similar to the normal Brillouin gain case. Thus, it can be supposed that k_2 and the GVD, respectively, have also the same frequency-characteristics for both cases. The real part of the second derivative of the complex wave number k_2 yields the GVD which is given by:

$$GVD \propto \Re(k_2 L) = \Re\left[\frac{d^2 k(\omega) L}{d\omega^2}\right] = g_1 \frac{\gamma_1 (\omega - \omega_0)^3 - 3\gamma_1^3 (\omega - \omega_0)}{\left[(\omega - \omega_0)^2 + \gamma_1^2\right]^3}$$
$$- g_2 \frac{\gamma_2 (\omega - \omega_0)^3 - 3\gamma_2^3 (\omega - \omega_0)}{\left[(\omega - \omega_0)^2 + \gamma_2^2\right]^3}.$$
(5.7)

In Figure 5.1d the GVD according to Equation (5.7) in comparison to the GVD of a normal Brillouin gain is shown. As can be seen, the GVD runs indeed equally with the frequency for both cases. Hence, it can be assumed that with this method the pulse broadening due to the delay process is not increased for the same time delay values.

5.2.2 Experimental Verification and Results

The experimental verification of this idea is described in [204] and can be seen in Figure 5.2. To produce the independent Brillouin gain and Brillouin loss two separate DFB-LD were combined by a 3-dB-coupler. The first pump wave at frequency f_{p1} creates the natural Brillouin gain at $f_{p1} - f_B$. The pump power and hence, the time delay can be controlled with an EDFA. To superimpose the gain with the Brillouin loss, the frequency of the second pump wave has to be tuned exactly to $f_{p2} = f_{p1} - 2f_B$. For the broadening of the loss the second pump laser was directly modulated with a noise signal. The pump power for the loss was controlled by a VOA. Inside the used SSMF with a length of 50 km the gain superimposes with the loss and the pulses are delayed. Therefore, the Gaussian-shaped pulses with an input pulse width of 35 ns at a wavelength of around 1550 nm were used. To prove the statements of the previous section, the time delay was measured depending on the gain pump power without and with the additional loss at different loss pump powers. Additionally, the overall Brillouin spectrum was

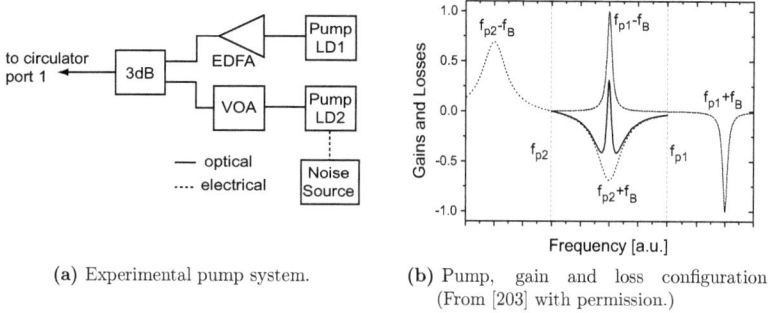

(a) Experimental pump system.

(b) Pump, gain and loss configuration. (From [203] with permission.)

Figure 5.2: Experimental verification for a time delay enhancement via superposition of a narrow Brillouin gain with a broadened Brillouin loss.

5.2 Superposition of a Narrow Brillouin Gain With a Broadband Brillouin Loss

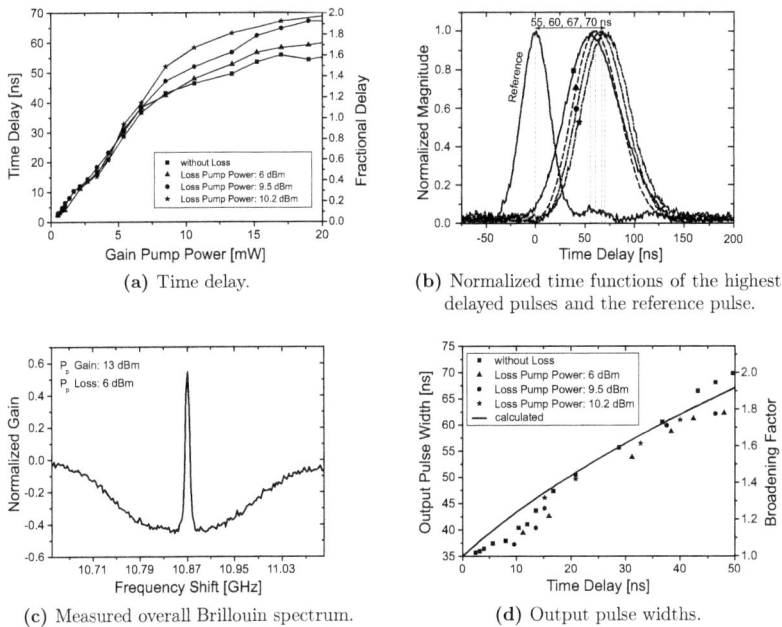

Figure 5.3: Measurement results for the proof of concept of a time delay enhancement via superposition of a narrow Brillouin gain with a broadened Brillouin loss.

measured as described in Section A.3. For the experiments the loss FWHM bandwidth was broadened to approximately 180 MHz.

The results of a first measurement using this method have been presented in [204] and are shown in Figure 5.3. The measured overall Brillouin spectrum for a loss pump power of 6 dBm and a gain pump power of 13 dBm can be seen in Figure 5.3c. Due to the superposition of the broad loss and the narrow gain the peak value of the spectrum is reduced. This depends on the relations between the optical power and the bandwidths of the pump spectra. Thereby, the loss maximum was only 40 % of that of a normal gain and the loss bandwidth of 180 MHz was five times higher than the natural gain bandwidth of 35 MHz. Within this gain bandwidth the signal pulses were delayed.

The time delay as a function of the gain pump power with and without the additional loss spectrum is shown in Figure 5.3a. Before the pump power saturates the delay increases equally for all conditions. Thus, the time delay does not change by adding the loss spectrum at first. For the case without the loss spectrum the saturation regime caused by the pump depletion is reached at approximately 35 ns which corresponds to $\Delta T_{frac} = 1$. However, by adding the loss spectrum the pump depletion is reduced. This reduction becomes significantly higher if the loss pump power is increased. As can be seen from the diagram, the saturation regime is reached at higher pump powers of the gain laser and therefore, at a higher time delay, for example at approximately 55 ns ($\Delta T_{frac} = 1.57$) for a loss pump power of 10.2 dBm. Thus, higher Brillouin losses result in higher achievable maximum time delays. Without the loss spectrum the maximum achieved time delay was only approximately 55 ns for a gain pump power of 20 mW whereas with the additional loss spectrum time delays of 60 ns, 67 ns and 70 ns have been obtained for loss pump powers of 6 dBm, 9.5 dBm and 10.2 dBm. Hence, the maximum fractional time delay of 1.57 of the normal gain case was enhanced by 27 % to $\Delta T_{frac} = 2$.

The time functions of all measured pulses can be found in Figure B.2 of Appendix B. The normalized time functions of the output pulses at the maximum time delay for all cases are shown in Figure 5.3b. From the diagram it can be assumed that the output pulse width is not further broadened due to the additional loss spectrum. This assumption is confirmed by the measurement of the output pulse width, as can be seen in Figure 5.3d. Although the determined values slightly deviate from the calculated square root fit, the progress of pulse widths is almost the same for all cases. The pulse width and hence, the broadening factor for the superimposed Brillouin spectrum is indeed not higher than that for a natural gain at the same time delays as assumed in Subsection 5.2.1. Thereby, the deviations between the theoretical and practical pulse broadening are caused by the averaging of the measured pulses and the neglect of higher-order dispersions for the calculated pulses.

In another measurement it has been shown how the time delay is decoupled from the Brillouin amplifier gain by this method and that for the same absolute gain the time delay is higher if an additional loss is used [205]. Thereby, the amplification gain was determined without considering the fiber attenuation. A comparison between the pulse delays with (dashed-dotted curves) and without an additional loss spectrum (dashed curves) for three different gains can be seen in Figure 5.4a. The reference signal (solid curve) is the pulse which can be measured at the system output if no pump signal is present. For a gain of 8.5 dB the pulse delay is around 37 ns when no additional loss spectrum is used. With such a loss spectrum the delay is increased to approximately 60 ns. Therefore, the gain pump power was increased so that the absolute gain remained the same while the group index is further increased due to the additional loss spectrum. For an absolute gain of around 13 dB time delays of 43 ns and 88 ns were achieved.

The highest time delay obtained with this method is nearly 100 ns which corresponds to three times the initial pulse width [203] which is shown in Figure 5.4b. If no loss spectrum is present the time delay at the same gain value of 16 dB is only 46 ns ($\Delta T_{frac} = 1.35$). Hence, using this method the time delay can be enhanced by approximately 122 %. If higher losses are used the group index and the time delay could be further enhanced. However, since for such high gains the system already works in the pump depletion regime this enhancement is limited. In principle, the maximum time delay depends on the Brillouin threshold of the gain and the loss. Therefore, the change of the group index which is induced by the loss can be almost doubled [203]. If the threshold for the Brillouin loss is increased a further enhancement would be possible. A method to realize this enhancement is proposed in Section 5.4.

For certain applications it might be useful that the amplitude of the pulse is not changed

5.2 Superposition of a Narrow Brillouin Gain With a Broadband Brillouin Loss

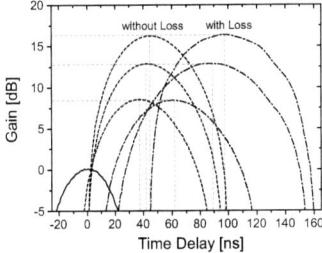

(a) Reference pulse and delayed pulses with and without additional loss spectrum for different amplifier gains (fiber attenuation neglected).

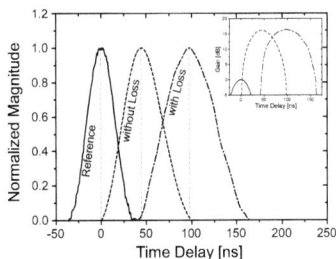

(b) Normalized time functions of the highest delayed pulses and the reference pulse. The inset shows the pulse magnitudes on a logarithmic scale without normalization.

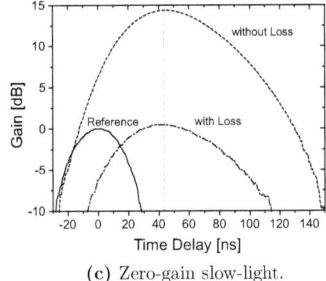

(c) Zero-gain slow-light.

Figure 5.4: Time delay enhancement and zero-gain slow-light via superposition of a narrow Brillouin gain with a broadened Brillouin loss.

during the delay process. This feature, also known as zero-gain slow-light, can also be applied by this superposition method of a narrow Brillouin gain with a broad Brillouin loss. In Figure 5.4c such a zero-gain pulse delay is shown. If the pulse is delayed in a conventional manner a maximum time of approximately 42 ns was achieved [205]. In this case the amplification is around 14 dB, which comes near the pump depletion regime. If the pulse is delayed with an additional loss spectrum the amplification can be reduced to 0 dB while the group index change and the time delay remains the same. Hence, with this method the delay process will not change the amplitude of the delayed pulse.

The approach of zero-gain slow-light was also used by other research groups. For example, Chin et al. also generated a synthesized Brillouin gain spectrum through the superposition of gain and loss profiles showing very different spectral widths [202]. By using a natural Brillouin gain and a loss with a bandwidth around 300 MHz they achieved an overall Brillouin profile very similar to an ideal EIT profile. Within the transparency range of this spectrum a time delay for 50 ns pulses was shown with minor amplitude changes. By using an inverse Brillouin profile it was also possible to advance the pulses without changing the pulse amplitudes.

Zhu et al. used two widely separated anti-Stokes absorption resonances in a HNLF to achieve nearly transparent slow-light [206]. Therefore, they created two pump lines by an external modulation of the pump laser. The frequency separation of these two pump lines defines the magnitude and the bandwidth of the occurring transparency window between the two Brillouin losses in the fiber. Within a 150 MHz broad transparency window 9 ns input pulses were delayed to $\Delta T_{frac} = 0.3$.

5.3 Superposition of a Brillouin Gain With Two Brillouin Losses at its Spectral Boundaries

A further opportunity to enhance the time delay and to achieve even higher delays as for the superposition of the narrow gain with a broadened loss is the superposition of a Brillouin gain with two losses at its spectral boundaries. Therefore, the theoretical background and experimental results are presented in this section. Furthermore, since this method provides several degrees of freedom for the adjustment of the slow-light performance, an optimization of different parameters, such as the frequency separation of the losses, is discussed.

5.3.1 Background Theory

For the superposition with two losses either a Brillouin gain with the natural bandwidth or a Brillouin gain with a broadened bandwidth can be chosen. Thus, the theoretical and mathematical background of both alternatives are slightly different.

Natural Brillouin Gain:

If a Lorentzian-shaped gain g_1 is superimposed with two Lorentzian-shaped losses g_2 at its spectral boundaries the complex wave number is [165]:

$$\begin{aligned} k(\omega) &= \frac{n}{c}\omega + \frac{1}{2L}\left\{\frac{g_1\gamma_1}{(\omega-\omega_0)+j\gamma_1} - \frac{g_2\gamma_2}{[\omega-(\omega_0\pm\delta)]+j\gamma_2}\right\} \\ &= \frac{n}{c}\omega + \frac{1}{2L}\left\{\frac{g_1\gamma_1(\omega-\omega_0)}{(\omega-\omega_0)^2+\gamma_1^2} - \frac{g_2\gamma_2[\omega-(\omega_0\pm\delta)]}{[\omega-(\omega_0\pm\delta)]^2+\gamma_2^2}\right. \\ &\quad -j\left[\frac{g_1\gamma_1^2}{(\omega-\omega_0)^2+\gamma_1^2} - \frac{g_2\gamma_2^2}{[\omega-(\omega_0\pm\delta)]^2+\gamma_2^2}\right]\right\}, \end{aligned} \quad (5.8)$$

with 2δ as the angular frequency separation between the two loss spectra. For this case the complex transfer function of the slow-light system becomes:

5.3 Superposition of a Brillouin Gain With Two Brillouin Losses

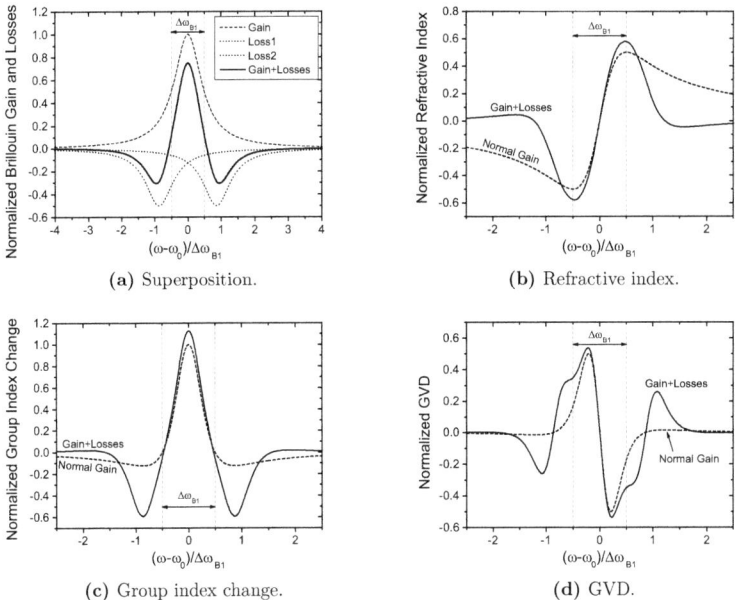

Figure 5.5: Behavior of a superposition (solid) of a natural gain (dashed) with two losses (dotted) at its spectral boundaries with $g_1 = 2g_2$ and $\Delta\omega_{B1} = \Delta\omega_{B2}$. All parameters are normalized to the natural Brillouin gain case.

$$H(\omega) = \exp\left\{\frac{g_1}{2}\frac{\gamma_1^2}{(\omega-\omega_0)^2+\gamma_1^2} - \frac{g_2}{2}\frac{\gamma_2^2}{[\omega-(\omega_0\pm\delta)]^2+\gamma_2^2}\right.$$
$$\left. + j\left[\frac{g_1}{2}\frac{\gamma_1(\omega-\omega_0)}{(\omega-\omega_0)^2+\gamma_1^2} - \frac{g_2}{2}\frac{\gamma_2[\omega-(\omega_0\pm\delta)]}{[\omega-(\omega_0\pm\delta)]^2+\gamma_2^2} + \frac{n}{c}\omega L\right]\right\}.$$
(5.9)

Identically to the previous derivations the frequency-distribution of the overall gain $g_B(\omega) = 2\Re[jk(\omega)L]$ and the change of the refractive index can be deduced from the real and the imaginary parts of the exponent in Equation (5.9). The overall Brillouin spectrum for a superposition of the gain and two losses with equal bandwidths and $g_1 = 2g_2$ is shown in Figure 5.5a. From the diagram it can be seen that the gain maximum is again decreased due to the losses. Hence, the time delay is decoupled from the amplifier gain and higher possible pump powers lead to an enhancement of the time delay because the pump depletion is reduced. This is the same behavior as for the method described in Section 5.2. In contrast to this, the gradient of the gain distribution is now changed as well. This leads to a higher slope of the refractive index alteration within the gain bandwidth, as can be seen in Figure 5.5b. Due to this higher SBS induced dispersion the time delay of the pulses will be increased.

The first derivative of $k(\omega)$ leads to the reciprocal group velocity:

$$\frac{dk(\omega)}{d\omega} = k_1 = \frac{1}{v_g} = \frac{n}{c} - \frac{1}{2L}\left\{\frac{g_1\gamma_1}{[(\omega-\omega_0)+j\gamma_1]^2} - \frac{g_2\gamma_2}{\{[\omega-(\omega_0\pm\delta)]+j\gamma_2\}^2}\right\}, \quad (5.10)$$

with the real part of k_1:

$$\Re(k_1) = \frac{n}{c} + \frac{1}{2L}\left\{g_1\frac{\gamma_1^3 - \gamma_1(\omega-\omega_0)^2}{\left[(\omega-\omega_0)^2+\gamma_1^2\right]^2} - g_2\frac{\gamma_2^3 - \gamma_2[\omega-(\omega_0\pm\delta)]^2}{\left\{[\omega-(\omega_0\pm\delta)]^2+\gamma_2^2\right\}^2}\right\}. \quad (5.11)$$

From Equation (5.11) it follows for the time delay in the line center of the gain distribution that:

$$\Delta T = \frac{g_1}{2\gamma_1} + g_2\gamma_2\frac{\delta^2 - \gamma_2^2}{(\delta^2 + \gamma_2^2)^2}. \quad (5.12)$$

As can be seen from Equation (5.12), the time delay additionally depends on the frequency separation of the two losses. As long as δ is higher than the loss bandwidths γ_2 the time delay is increased by the loss spectra at the spectral boundaries of the Brillouin gain. For the superposition in Figure 5.5 a frequency separation of the losses from the line center of $\delta = \gamma_2\sqrt{3}$ was used. In Figure 5.5c the normalized group index as a function of the normalized frequency is shown. As can be seen from the diagram, the line center group index is indeed increased although the peak value of the gain distribution is reduced. Thus, the presence of the additional losses leads directly to a time delay enhancement in contrast to the superposition of a narrow Brillouin gain with a broadened Brillouin loss where the loss decreases the time delay.

With $\delta = \gamma_2\sqrt{3}$ the slow-light delay in the line center becomes:

$$\Delta T = \frac{g_1}{2\gamma_1} + \frac{g_2}{8\gamma_2} = \frac{g_1}{2\pi\,\Delta f_{B1}} + \frac{g_2}{8\pi\,\Delta f_{B2}}. \quad (5.13)$$

If the gain and the losses have equal bandwidths, i.e. $\gamma_1 = \gamma_2 = \gamma$, the time delay is:

$$\Delta T = \frac{1}{2\gamma}\left(g_1 + \frac{1}{4}g_2\right) = \frac{1}{2\pi\,\Delta f_B}\left(g_1 + \frac{1}{4}g_2\right). \quad (5.14)$$

At the same time the maximum of the Brillouin gain in the line center under these conditions becomes:

$$g_B(\omega_0) = g_1 - \frac{1}{2}g_2. \quad (5.15)$$

If the gain and the losses have the same magnitudes, i.e. $g_1 = g_2 = g$, the time delay and the peak value of the gain yield:

$$\Delta T = \frac{5g}{8\gamma} \quad \text{and} \quad g_B(\omega_0) = \frac{1}{2}g. \quad (5.16)$$

In comparison to a basic SBS-based slow-light system with a natural single Brillouin gain ($\Delta T = g/(2\gamma)$) the time delay is enhanced by 25 %. This improvement is caused by the increased gradient of the gain distribution. Additionally, the time delay scales with the amplifier gain and the pump power, respectively. As can be seen from Equation (5.16) the two losses decreases the gain maximum which is only 50 % compared to the gain maximum of the natural single gain case. Thus, $g_B(\omega_0)$ can be doubled by using higher pump powers and the time delay enhancment due to the delay-gain-decoupling is 100 %. Hence, the entire improvement of the time delay is 125 % under these circumstances. According to Equation (5.14), the time delay can be further enhanced if the powers of the anti-Stokes resonances are increased.

The GVD of the superposition of a Brillouin gain with two losses at its spectral boundaries, deduced from the second derivative of $k(\omega)$, is described by:

$$GVD \propto \Re(k_2 L) = \Re\left[\frac{d^2 k(\omega) L}{d\omega^2}\right] = g_1 \frac{\gamma_1(\omega-\omega_0)^3 - 3\gamma_1^3(\omega-\omega_0)}{\left[(\omega-\omega_0)^2 + \gamma_1^2\right]^3}$$

$$-g_2 \frac{\gamma_2[\omega-(\omega_0 \pm \delta)]^3 - 3\gamma_2^3[\omega-(\omega_0 \pm \delta)]}{\left\{[\omega-(\omega_0 \pm \delta)]^2 + \gamma_2^2\right\}^3}.$$

(5.17)

The normalized GVD as a function of the normalized frequency according to Equation (5.17) is shown in Figure 5.5d. As can be seen from the diagram, the GVD within the gain bandwidth is slightly increased by the additional losses. Hence, this higher slope will lead to a higher temporal broadening of the delayed pulses in comparison to a single gain slow-light system. Due to the higher output pulse width the effective time delay will be decreased. However, since the time delay enhancement is very high the effective time delay for this method can nevertheless be higher than that for the single gain case, as will be shown in Subsection 5.3.3.

Broadened Brillouin Gain:

For applications in optical communications, such as optical buffers or synchronizers, the natural Brillouin bandwidth is too low and has to be increased. This can be realized by the methods described in Chapter 4. In comparison to the Brillouin bandwidth enhancement via a polychromatic pump source, using a broadband pump provides extremely wideband Brillouin gains. Thus, only this method is further considered in this section.

If the Brillouin gain is enhanced by a broadband pump its spectrum has a Gaussian distribution. Hence, the complex wave number for a broad gain spectrum is given approximately by [165, 195]:

$$k(\omega) = \frac{n}{c}\omega + \frac{g_1}{j2L} e^{-\left(\frac{\omega-\omega_0}{\gamma_G}\right)^2} \mathrm{erfc}\left(-j\frac{\omega-\omega_0}{\gamma_G}\right),$$

(5.18)

with γ_G as the half $1/e$-bandwidth of the gain distribution and $erfc()$ as the complementary error function. The complex error function cannot be solved analytically. Hence, the gain spectrum, the frequency-distribution of n as well as the derivatives of $k(\omega)$ were determined by using the algebra software Maple version 11.

The time delay in the line center of a Gaussian-shaped Brillouin gain is expressed by:

$$\Delta T = \frac{g_1}{\sqrt{\pi}\,\gamma_G} \approx 0.47\frac{g_1}{\gamma_1} \quad \text{with} \quad \gamma_1 = \gamma_G\sqrt{ln(2)}. \tag{5.19}$$

As can be seen, this equation only slightly differs from Equation (3.40), the line center time delay for a Lorentzian-shaped gain. Therefore, the shape of the Brillouin gain has a minor influence on the slow-light delay. The time delay is strongly decreased for both shapes if the bandwidth is increased for equal gain values.

If Equation (5.18) is introduced into Equation (5.8) the complex wave number for a superposition of a broadened Brillouin gain with two losses becomes:

$$k(\omega) = \frac{n}{c}\omega + \frac{1}{2L}\left\{\frac{g_1}{j}e^{-\left(\frac{\omega-\omega_0}{\gamma_G}\right)^2}erfc\left(-j\frac{\omega-\omega_0}{\gamma_G}\right) - \frac{g_2\gamma_2}{[\omega-(\omega_0\pm\delta)]+j\gamma_2}\right\}. \tag{5.20}$$

Hence, the complex transfer function for this method can be written as:

$$\begin{aligned}H(\omega) = {} & \exp\left\{\frac{g_1}{2}e^{-\left(\frac{\omega-\omega_0}{\gamma_G}\right)^2}erfc\left(-j\frac{\omega-\omega_0}{\gamma_G}\right)\right.\\ & -\frac{g_2}{2}\frac{\gamma_2^2}{[\omega-(\omega_0\pm\delta)]^2+\gamma_2^2} \\ & \left. + j\left[\frac{n}{c}\omega L - \frac{g_2}{2}\frac{\gamma_2\,[\omega-(\omega_0\pm\delta)]}{[\omega-(\omega_0\pm\delta)]^2+\gamma_2^2}\right]\right\},\end{aligned} \tag{5.21}$$

and the time delay yields:

$$\Delta T = \frac{g_1}{\sqrt{\pi}\,\gamma_G} + g_2\gamma_2\frac{\delta^2-\gamma_2^2}{(\delta^2+\gamma_2^2)^2} \approx 0.47\frac{g_1}{\gamma_1} + g_2\gamma_2\frac{\delta^2-\gamma_2^2}{(\delta^2+\gamma_2^2)^2}. \tag{5.22}$$

Also this equation differs only slightly from the Lorentzian-shaped gain case. Hence, for this method the same conditions are valid.

The behavior of the superposition of a broadened Brillouin gain with two losses at its spectral boundaries in comparison to a natural single gain is shown in Figure 5.6. For the diagrams equal magnitudes of the gain and the losses were chosen and the gain bandwidth was broadened to three times the loss bandwidth. The separation of the losses from the line center is $\delta = \gamma_{B2}\sqrt{3}$. As can be seen from Figure 5.6a, the base line of the gain is shifted into the loss region. Thus, the maximum value of the gain as well as the pump depletion is reduced and the time delay is decoupled from the amplifier gain again. This leads to a time delay enhancement. However, due to the spectral broadening the slope of the refractive index within the gain bandwidth and therefore, the group index change in the line center is decreased. By adding the additional losses at the spectral boundaries of the broadened gain the refractive index slope and hence, the line center group index change is increased as in the case of a natural Brillouin gain superimposed

5.3 Superposition of a Brillouin Gain With Two Brillouin Losses

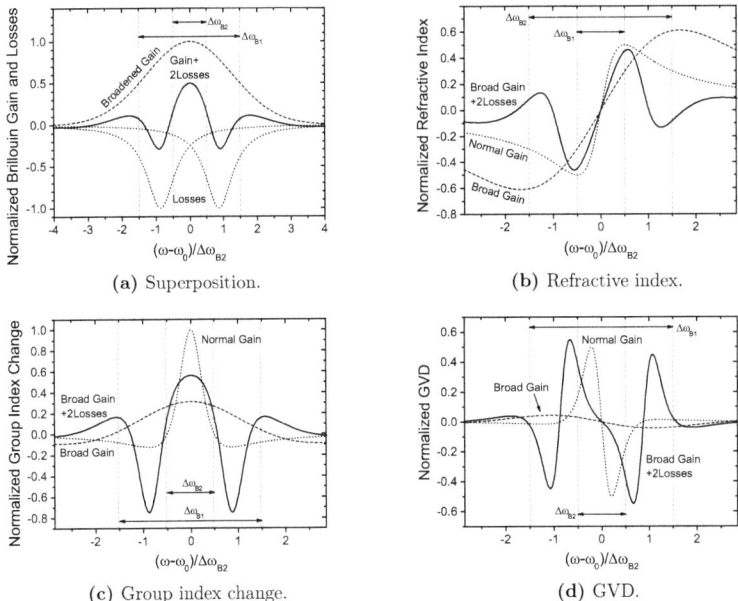

Figure 5.6: Behavior of a superposition (solid) of a broadened gain with two losses at its spectral boundaries in comparison to a single normal (dotted) and a broadened gain (dashed) with $g_1 = g_2$, $\Delta\omega_{B1} = 3\Delta\omega_{B2}$ and $\delta = \gamma_{B2}\sqrt{3}$. All parameters are normalized to the natural Brillouin gain case.

with two losses. Nevertheless, both parameters are still lower than for a single natural Brillouin gain. However, due to the delay-gain-decoupling this can be further enhanced. Additionally, the broadening of the gain leads to a much lower GVD, as can be seen in Figure 5.6d. Thus, in summary, a smaller time delay on the one hand but also a smaller temporal pulse broadening on the other hand can be expected in comparison to a slow-light system with a single natural Brillouin gain.

5.3.2 Optimization

The tailoring of the Brillouin spectrum by the superposition of a Brillouin gain with two losses at its spectral boundaries offers several degrees of freedom:

1. The magnitude of the losses g_2 in relation to the magnitude of the gain g_1 defines the amount of time delay enhancement due to the delay-gain-decoupling.

2. The frequency separation of the losses 2δ has a large impact on the slow-light bandwidth on the one hand. On the other hand, the position of the losses defines the increase of the gradient of the gain distribution and hence, time delay enhancement.

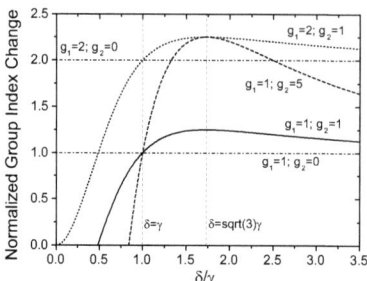

Figure 5.7: Normalized line center group index change as a function of the normalized frequency separation of the loss spectra for different relations between the gain and the losses with $\gamma_1 = \gamma_2$. All cases are normalized to the single natural gain case.

3. The bandwidth of the gain $\Delta\omega_{B1}$ defines the maximum delayable pulse width and data rate, respectively. Furthermore, the bandwidth has a significant impact on the pulse distortions.

Hence, this superposition method offers the opportunity to adapt all properties of the Brillouin spectrum, such as the bandwidth, slope and gain, to the given application [207, 208].

The normalized group index change at $\omega = \omega_0$ as a function of the frequency separation between the loss spectra for different relations between the gain and the losses and equal gain and loss bandwidths can be seen in Figure 5.7. If $\delta/\gamma < 1$ the losses reduce the gain drastically and hence, the time delay is smaller than without the additional losses. For $\delta/\gamma = 1$ the time delay is the same in both cases. If the frequency separation is higher the loss spectra will increase the time delay. This is due to the fact that the time delay is a function of the slope of the Brillouin gain and the additional loss spectra increase its gradient. Therefore, the maximum time delay can be found at $\delta = \gamma\sqrt{3}$ [165]. The enhancement of the time delay depends on both the loss and the gain power. If the gain is increased the time delay increases as well. Due to the decoupling of the time delay from the amplification process the relative gain can be much higher than without the additional losses. However, the absolute gain maximum remains constant and saturates the time delay for high pump powers. A further enhancement of time delay can be achieved if the loss power is increased.

Since the separation of the losses defines also the slow-light bandwidth additionally to the time delay enhancement, there might be different optimal operating points concerning these two parameters [209, 210]. In Figure 5.8 a comparison of the optimization between the superposition of the losses with a natural (on the left hand side) and a broadened Brillouin gain (on the right hand side) is shown. Thereby, the Brillouin gain was broadened to approximately three times of the loss bandwidth and equal powers for gain and losses were assumed for both cases. The contour plots show the frequency-distribution of the normalized group index change as a function of the loss separation. From these plots the normalized time delay in the line center, which is proportional to the group index, and the FWHM bandwidth of the group index change

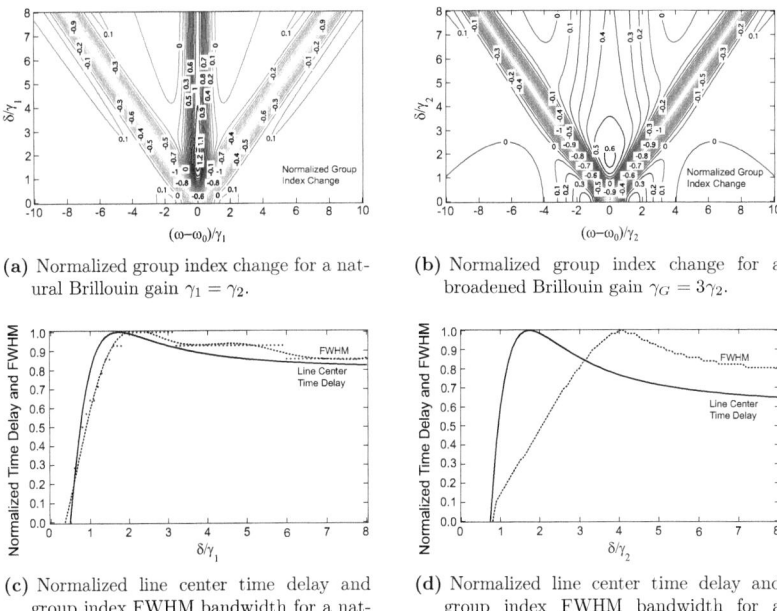

Figure 5.8: Optimization of the normalized group index change, line center time delay and FWHM bandwidth by the variation of the loss separation for a superposition of a Brillouin gain with two losses at its spectral boundaries for $g_1 = g_2$.

as a function of δ were extracted. The discrete values of the FWHM bandwidth have been fitted to a 9th order polynom.

As already explained and as can be seen from Figure 5.8c the maximum time delay for the superposition with the normal gain is achieved at $\delta = \gamma_2\sqrt{3} \approx 1.7\gamma_2$. For a natural Brillouin bandwidth of 35 MHz this corresponds to $\delta/(2\pi) \approx 30$ MHz. At the same time the maximum bandwidth is also nearby this position. The reason for the small drift is that the losses only marginally overlap with the gain. Furthermore, the interpolation of the curve might add a small deviation. However, the bandwidth of the group index change quantifies the temporal pulse broadening which is minimal for the maximum bandwidth. Thus, the optimal point for the maximum effective time delay can be expected in the region of $\delta = \gamma_2\sqrt{3}$ as well. This means that for the superposition of a natural gain with two losses only one optimal operating point exists.

If the Brillouin gain is broadened a higher slow-light bandwidth is available. Hence, the optimal point for the highest bandwidth will be increased to higher values of δ. As can be seen in Figure 5.8d, this occurs at $\delta \approx 4\gamma_2$. However, at this point less than 80 % of the maximum time delay can be achieved. As for the natural gain case, the maximum time delay is obtained at $\delta = \gamma_2\sqrt{3}$. For this frequency separation the losses are very close to the line center of the gain which therefore, leads to a significant reduction of the gain bandwidth. Thus, the output pulse widths of the delayed pulses can be very high. Therefore, the optimal point for the maximum effective time delay is shifted and can be expected in the range of $1.7\gamma_2 < \delta < 4\gamma_2$. Hence, such a system configuration offers three different operating points for the properties of a high time delay, a high bandwidth (low distortion) and a trade-off between both conditions.

5.3.3 Experimental Verification and Results

To achieve the superposition of the Brillouin gain with two losses at its spectral boundaries the experimental pump system setup and the frequency configuration of the pumps was used according to Figure 5.9. Similar to the setup described in Subsection 5.2.2, the pump waves of two independent DFB-LD were combined with a frequency separation of $f_{p1} = f_{p2} + 2f_B$ by a 3-dB-coupler. The pump power for the gain and the losses are adjusted by two EDFA. For the spectral broadening of the natural Brillouin gain the first pump wave was directly modulated by a Gaussian noise signal. The two losses were produced by an external modulation of the second pump source via a MZM. Therefore, the output power of the modulator was maximized by adjusting the OPC. The bias voltage was set to drive the MZM in the upper quadratic operating point. Thus, a carrier suppression is achieved and both first-order sidebands are created. The frequency separation of the losses depends on the sine frequency of the modulation signal and is therefore $2f_{mod}$. Inside the slow-light medium (50 km of SSMF) the gain and the two losses superimpose at $f_{p1} - f_B = f_{p2} + f_B$ where the 30 ns pulses are delayed.

At first, the optimization of the superposition of a natural and a threefold broadened gain with the two losses is investigated [209, 210]. The time delay and the pulse widths were measured in dependence of the separation of the losses. In Figure 5.10 the normalized time delays and normalized pulse widths as a function of the frequency separation of losses are shown. Thereby, both parameters are normalized to their maximum measured values. Additionally, the relation

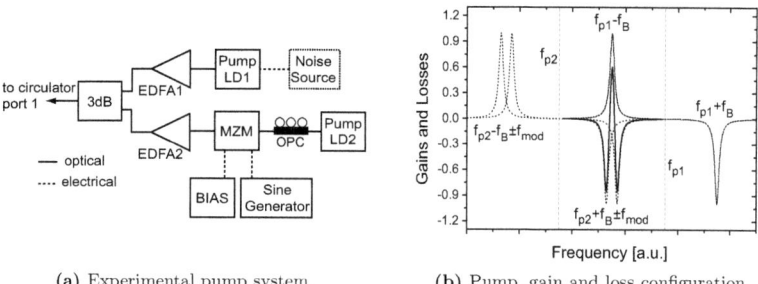

(a) Experimental pump system.   (b) Pump, gain and loss configuration.

Figure 5.9: Experimental verification for a time delay enhancement via superposition of a Brillouin gain with two Brillouin losses.

5.3 Superposition of a Brillouin Gain With Two Brillouin Losses

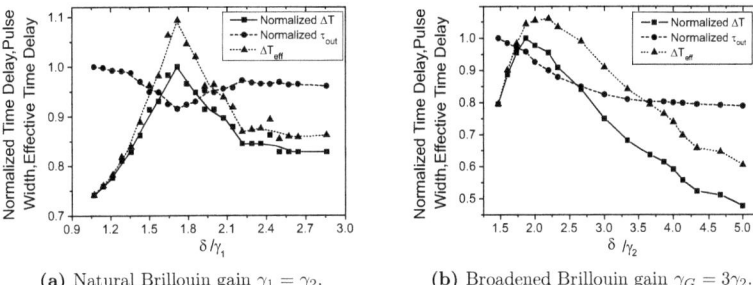

(a) Natural Brillouin gain $\gamma_1 = \gamma_2$. (b) Broadened Brillouin gain $\gamma_G = 3\gamma_2$.

Figure 5.10: Optimization of the normalized time delay, output pulse width and effective time delay as a function of the loss separation for the superposition of a Brillouin gain with two Brillouin losses at its spectral boundaries. The time delay and pulse widths were normalized to its maximum values and the effective time delay is the relation between these two values.

of the normalized time delay and pulse width was determined to get the effective time delay.

The measurement for equal natural bandwidths of the gain and the losses is shown in Figure 5.10a. The pump power of the losses and the gain were 10 dBm and 10.5 dBm. With an increasing separation the time delay increases as well up to a maximum value at $\delta \approx 1.7\gamma_1$. At the same time the Brillouin bandwidth increases up to a maximum value. Among other parameters the slow-light bandwidth directly defines the temporal broadening of the pulses. Therefore, the pulse width decreases to a minimum due to the increasing bandwidth. The optimal point for this can also be found at $\delta \approx 1.7\gamma_1$. The reason for this is that the spectrum for the 30 ns wide Gaussian-shaped input pulses is only approximately 15 MHz. Hence, for a loss separation of $2\delta \approx 3.4\gamma_1$, which corresponds to a modulation frequency $f_{mod} \approx 30$ MHz, the losses do not restrict the amplified pulse spectrum leading to the minimum output pulse width. Therefore, the maximum effective time delay can be found in the same region of the maximum time delay and minimum pulse width, respectively. With a further increase of the separation the influence of the losses decreases and hence, the time delay diminishes. Thus, as supposed, for the superposition of a natural Brillouin gain and two losses with equal bandwidths only one optimal operating point exists.

If the gain spectrum is broadened – in this case by a factor of three – the three optimal points occur at different values of δ, as can be seen in Figure 5.10b. The gain pump power had to be increased to 15 dBm and the loss pump power was adapted to 11 dBm. The maximum time delay was achieved at $\delta \approx 1.8\gamma_2$ and is near the value of the previous measurement. The pulse width decreases with higher separation of the losses and approaches a minimum value because of the higher bandwidth. Hence, the maximum effective time delay value is shifted to $\delta \approx 2.25\gamma_2$. For wider gain bandwidths it can be expected that this optimal point would be further shifted to higher values of δ. As expected from the theory, an optimization of such a slow-light system for different properties is possible by choosing three available operating

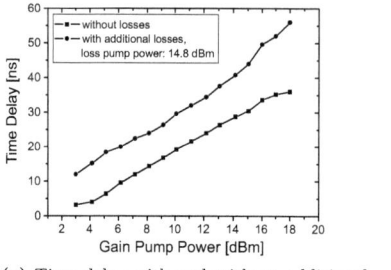
(a) Time delay with and without additional loss spectra.

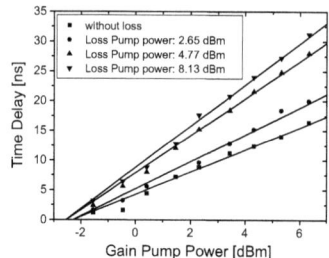
(b) Time delay for different loss pump powers.

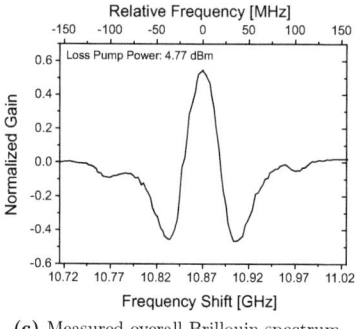
(c) Measured overall Brillouin spectrum.

Figure 5.11: Measurement results for the proof of concept of a time delay enhancement via superposition of a Brillouin gain with two losses at its spectral boundaries.

points: the maximum time delay, the minimum pulse width and the maximum effective time delay which is a trade-off between the time delay and the bandwidth.

In various proof of concept measurements it was shown that the superposition of a natural Brillouin gain with two losses at its spectral boundaries leads to an enhancement of the time delay [207, 211]. In Figure 5.11a the result of a first measurement is shown. As can be seen from the diagram, if no additional loss spectra are present the time delay has a maximum of around 35 ns. If the pump power is further increased the pulse delay is saturated. However, by adding the two loss spectra ($\delta = \gamma_1\sqrt{3}$, $P_{p2} = 14.8$ dBm) the gradient of the Brillouin spectrum is increased. This leads to a higher group index change and therefore, to a higher maximum time delay of 57 ns for the same gain pump power. Due to the fact that the absolute gain is reduced as well higher pump powers can be used until the time delay is saturated. An example of a measured overall Brillouin spectrum for this method is shown in Figure 5.11c. It can be clearly recognized that the two losses are placed at the spectral boundaries of the gain.

If the losses are increased the time delay can be further enhanced. This was investigated in another measurement. The time delay as a function of the gain pump power for four different

5.3 Superposition of a Brillouin Gain With Two Brillouin Losses

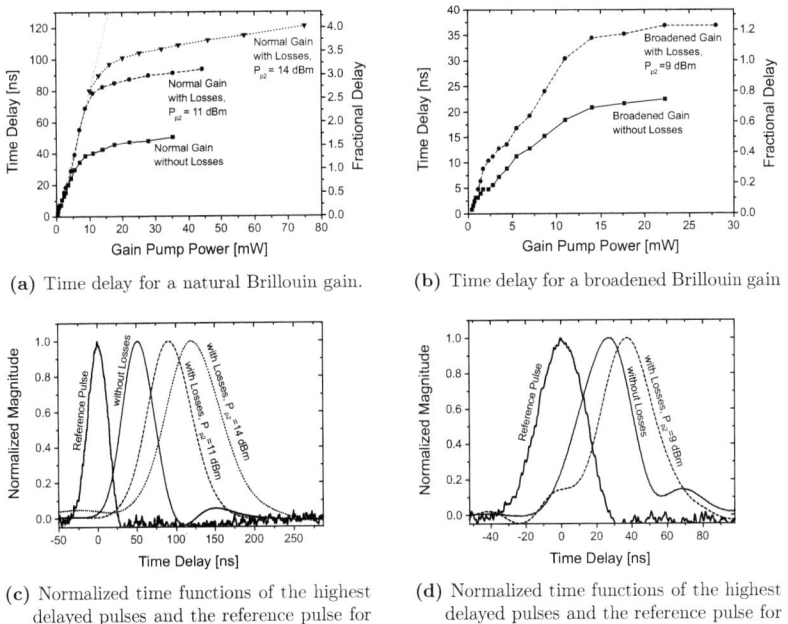

Figure 5.12: Measurement results for a time delay enhancement via superposition of a natural and broadened Brillouin gain with two Brillouin losses at its spectral boundaries. The loss separation is approximately 60 MHz for both cases.

powers of the losses can be seen in Figure 5.11b. Without the losses only the Brillouin gain is present which leads to a pulse delay of approximately 16 ns for a gain pump power of 6.3 dBm. If the two loss spectra are introduced with a rather low power of 2.65 dBm for the same gain pump power the delay is increased to 20 ns. By generating the two loss spectra with a high power of 8.13 dBm the delay reaches a value of around 31 ns, which is almost twice of that without a loss.

The results for a further measurement on this method can be seen in Figure 5.12. The diagrams on the left hand side show the results for a superposition of a Brillouin gain and two losses with natural bandwidths of approximately 35 MHz. Without an additional loss the linear increase of the time delay with increasing pump power rises to approximately 30 ns. After that the time delay can be increased further but, due to the pump depletion the enhancement is rather low. If the two loss spectra are introduced with a frequency separation from the line center of approximately 30 MHz ($\delta = \gamma\sqrt{3}$) the linear time delay rises to around 70 ns and 80 ns for loss pump powers of 11 dBm and 14 dBm, respectively. If the gain pump power is increased further a time delay of more than 120 ns can be achieved in the saturated regime, as can be seen in Figure 5.12a. This corresponds to a fractional delay of four. Up to now this is the highest reported time delay in a SBS-based slow-light system with just one fiber spool [165]. However, due to the fact that the loss amplifier was restricted to 14 dBm this time delay is not the achievable maximum. Additionally, the temporal width of the output pulse was broadened to 85 ns, as can be seen in Figure 5.12c. Thus, the effective time delay is strongly reduced to 1.4.

To verify that this time delay enhancement method also works with higher bandwidths the gain was broadened to approximately 60 MHz. Due to the broadening the time delays are reduced, as shown in Figure 5.12b. For a gain pump power of 22 mW a time delay of around 23 ns was achieved if only the gain is present. In this case, the 30 ns input pulse was broadened to approximately 34 ns, as can be seen in Figure 5.12d. If the two loss spectra are introduced at the spectral boundaries of the gain with a frequency separation of again approximately 60 MHz and a loss pump power of 9 dBm the time delay is increased to 36 ns for the same gain pump power. Although this is a more than 50 % higher time delay the width of the output pulse was not further increased and amounts also to approximately 34 ns. Hence, as expected from the theory, this method can reduce the temporal pulse broadening. However, due to the measurement near to the sturation regime the delayed pulses shown in Figure 5.12d experience additional distortions of the pulse shape. The time functions of all measured pulses for this method can be found in Section B.3 and Section B.4.

5.4 Suppression of Spurious Backscattered Stokes-Waves in Multiple-Pump-Line Systems

As shown in Section 5.2 and Section 5.3 the time delay is enhanced if a Brillouin gain is superimposed with one broad or two narrow loss spectra. Therefore, if the loss power is increased the time delay enhancement can be increased. However, this leads to a creation of a backscattered spurious Stokes-wave which depletes the loss pump wave. Hence, the time delay enhancement is limited. To circumvent this limitation a new slow-light delay line configuration has been investigated which partially reduces the pump depletion for the losses due to a blocking of the spurious backscattered Stokes-wave [212, 213]. In this section, this novel method is examined on the superposition of a Brillouin gain with two losses.

5.4.1 Background Theory

If the frequencies ω and the HWHM bandwidths γ in Equation (5.8) or Equation (5.9) are normalized to the HWHM gain bandwidth γ_1 and the induced loss g_2 is normalized to the induced gain g_1 the overall Brillouin spectrum of the superposition of a gain with two losses at

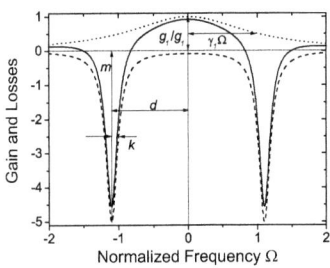
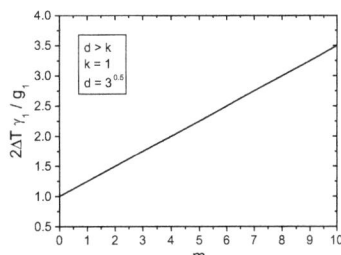

(a) Superposition with normalized parameters $m = g_2/g_1$, $k = \gamma_2/\gamma_1$, $d = \delta/\gamma_1$.

(b) Time delay as a function of the ratio between the losses and the gain.

Figure 5.13: Overall gain and time delay due to a superposition (solid) of a Brillouin gain (dotted) with two Brillouin losses (dashed) at its spectral boundaries.

its spectral boundaries can be written as [213, 214]:

$$g_B(\omega) = g_1 \left(\frac{1}{1+\Omega^2} - mk^2 \left[\frac{1}{k^2 + (\Omega+d)^2} + \frac{1}{k^2 + (\Omega-d)^2} \right] \right), \tag{5.23}$$

with $\Omega = (\omega - \omega_0)/\gamma_1$, $k = \gamma_2/\gamma_1$, $m = g_2/g_1$ and $d = \delta/\gamma_1$. The overall gain with these normalized parameters is shown in Figure 5.13a. By controlling the loss-gain-ratio m, the bandwidth ratio k and the separation of the two losses $2d$, the system can be adjusted to different properties. Then, the corresponding time delay at $\omega = \omega_0$ (according to Equation (5.12)) can be rewritten as:

$$\Delta T = \frac{g_1}{2\gamma_1}\left[1 + 2mk\frac{d^2 - k^2}{(d^2 + k^2)^2}\right]. \tag{5.24}$$

Equation (5.23) and Equation (5.24) are rather valid for Lorentzian-shaped gains. However, for the sake of simplicity they can also be applied for Gaussian-shaped gains in this case because the deviation from the equation for the line center time delay with Gaussian-shaped gains (Equation (5.22)) is marginal. It follows from Equation (5.24) that the influence of the losses will increase the time delay as long as $d > k$. If $k = 1$ and $d = k\sqrt{3}$ the maximum achievable time delay becomes:

$$\Delta T = \frac{g_1}{2\gamma_1}\left(1 + \frac{1}{4}m\right). \tag{5.25}$$

Hence, the time delay could be further enhanced by an increase of m, as can be seen in Figure 5.13b. Therefore, the loss pump power has to be enlarged. If it exceeds the Brillouin threshold the loss begins to saturate because the pump power is transferred to an exponentially rising backscattered Stokes-wave, as illustrated on top of Figure 5.14a. This spurious Stokes-wave restricts or even decreases the maximum achievable amount of the losses [164].

To overcome this drawback, the first idea is to use shorter fibers where the nonlinear interactive length is smaller for the losses. Hence, the Brillouin threshold is increased and the

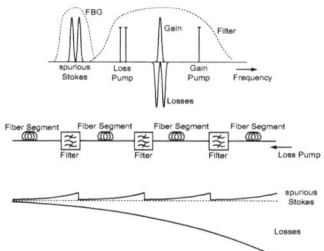
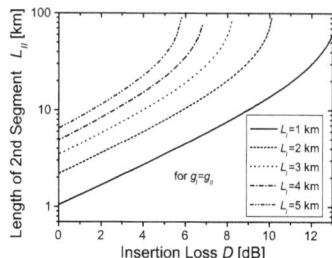

(a) Idea of spurious Stokes-wave blocking by a filter and a FBG (top). Slow-light medium configuration consisting of fiber segments with filters in between them (middle). Distribution of the losses and the spurious Stokes-wave along the segments (bottom).

(b) Length of the second fiber segment as a function of the insertion loss between two connected fibers to achieve equal gains if the first fiber length is given.

Figure 5.14: Suppression of spurious backscattered Stokes-waves and compensation of the insertion loss of the filter stages.

revoked pump power will be decreased, as described in Section 3.1. However, a shorter fiber length would also lead to a smaller gain and thus to a reduced time delay. Therefore, the second idea is to prevent the propagation of the spurious Stokes-wave generated by the losses along the optical waveguide by a filter. Thereby, the filter should not impair the gain, the losses and their pump waves. Hence, a long fiber divided into many short segments combined with a blocking of the spurious Stokes-wave, as shown in the middle of Figure 5.14a), would lead to an enhancement of the losses and the loss-gain-ratio. The result is that only the short fiber segments are present for the small spurious Stokes-waves generated by the loss pumps. Additionally, the spurious Stokes-waves will be canceled at every filter stage, whereas the configuration works like a long fiber for the Brillouin losses and the gain. They will increase exponentially along the whole length of all segments, as can be seen on the bottom of Figure 5.14a.

The blocking of the spurious Stokes-wave could be achieved by using the reflection characteristic of a FBG or a common WDM bandpass filter to transmit all designated spectral components (top of Figure 5.14a). However, as the measurement results will show, common filters do not provide optimal blocking. Due to their low slew rate it is not possible to separate the spectral components with a distance of exactly 11 GHz. This could be achieved by using customized filters with an improved slew rate.

Another problem is that the pump powers for the losses and the gain after the first segment are strongly reduced by the insertion loss of the filters. However, the main advantage of the proposed new slow-light configuration is that the fiber lengths can be adapted to provide the same gains. Therefore, the attenuation between the segments and the attenuation of the first fiber can be compensated. In the small pump power regime, the gains of two segments can be expressed by [213]:

5.4 Suppression of Spurious Backscattered Stokes-Waves in Multiple-Pump-Line Systems

and

$$g_I = \frac{g_B P_{p1} L_{eff,I}}{A_{eff}} \tag{5.26}$$

$$g_{II} = \frac{g_B P_{p1} \exp(-\alpha L_I) L_{eff,II}}{A_{eff} D} \tag{5.27}$$

with

$$L_{eff,I,II} = \frac{1 - \exp(-\alpha L_{I,II})}{\alpha}, \tag{5.28}$$

where α is the fiber attenuation constant in km^{-1} and D is the linear insertion loss of the filter between the segments. Thereby, the pump power of the second segment $P_{p1} \exp(-\alpha L_I)/D$ with the length L_{II} is reduced by D and the attenuation of the first fiber with the length L_I. In Figure 5.14b the length of the second segment is shown which is necessary to compensate the insertion loss of the filters and to get equal gains in both segments if input fiber lengths of 1 to 5 km ($\alpha = 0.2\,\mathrm{km}^{-1}$) are used. As can be seen from the diagram, the interconnection of fibers with a length of 2 km and 25 km can for instance compensate an insertion loss of 8.5 dB.

5.4.2 Experimental Verification and Results

For the experimental verification of this method the general setup described in Subsection 3.2.2 together with the pump setup of Subsection 5.3.3 were used. The suppression of the spurious backscattered Stokes-wave due to the losses was achieved with three novel configurations of the slow-light medium, as shown in Figure 5.15.

At first, two segments of SSMF with lengths of 5 km were used together with a tunable FBG (TFBG) with a FWHM reflection bandwidth of 45 GHz and a -3/-20 dB-bandwidth ratio of 0.6. The grating is adjusted to prevent the propagation of the backscattered Stokes-wave due to the high pump power of the losses. However, the low slew rate enabled only a partial blocking of the spurious wave because the TFBG could not be centered on the Stokes-wave's wavelength. Otherwise, the pump wave for the Brillouin losses would also be blocked. For the second configuration, the slow-light medium with the same adjustments was extended to three SSMF segments with lengths of 5 km and with two identical TFBG in between them.

As third configuration two fiber segments were used with a WDM filter with a FWHM bandwidth of 21 GHz and a -3/-20 dB-bandwidth ratio of 0.4 in between them. In contrast to the setup with the TFBG, different fiber lengths are necessary because the filter has a large insertion loss of 5.5 dB. Additionally, the center wavelength of the pulse was adjusted to the line center of the filter. Hence, the pump wavelengths coincide with the FWHM filter bandwidth and the pump waves were attenuated by a further 3 dB. To compensate the complete insertion

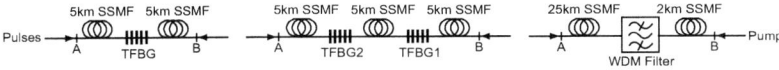

Figure 5.15: Slow-light medium configurations for spurious Stokes-wave suppression.

loss of 8.5 dB a SSMF of 2 km length was interconnected with a SSMF of 25 km length. Thus, the gains and losses of the first fiber segment are equal to gains and losses of the second fiber segment while m is maintained.

In comparison to cascaded systems [184, 196], this configuration is much easier because the fibers are only connected by the filters. All segments are pumped once by the same source like a passive optical delay line. Hence, there is no additional increase in noise. In contrast to this, in cascaded systems every segment is pumped separately by its own source. Therefore, more components are necessary and the setup is more complicated. Furthermore, the noise is increased in every cascade. However, as in cascaded systems a gain saturation coming from large amplification of the probe pulse also cannot be avoided with this new configuration. Contrary to the systems presented in [184, 196], here, the use of the additional losses at the spectral boundaries of the gain leads to an enhancement of the time delay, as previously described.

The measurements were performed with natural Brillouin bandwidth of approximately 35 MHz. The separation of the two losses was adjusted to 30 MHz ($\delta \approx \gamma\sqrt{3}$) to achieve the maximum time delay for the Gaussian-shaped input pulses with a temporal FWHM width of 30 ns. The measurement results for time delay enhancement via suppression of the backscattered spurious Stokes-wave caused by the loss spectra are shown in Figure 5.16. The diagrams on the left hand side show the fractional time delay, broadening factor and normalized time functions of selected pulses for the blocking with TFBG while the diagrams on the right hand side show the results for the blocking with the filter. The normalized time functions of all pulses for both cases can be found in Section B.5.

For the measurement with one TFBG and two 5 km SSMF segments the loss power was fixed to 18 dBm and the gain pump power was varied from 15 dBm to 22 dBm. As can be seen in Figure 5.16a and Figure 5.16c, a fractional time delay from 1.75 to 2.6 and a pulse broadening from 1.65 to 1.9 were achieved without a blocking of the Stokes-wave. However, by using the TFBG to jam this spurious signal the fractional time delay is enhanced by 8 % on average whereas the broadening factor is increased by just 4 %.

Almost the same time delay enhancement and pulse broadening increase can be obtained with two TFBG and three SSMF with lengths of 5 km. However, with this setup it is possible to increase the fixed loss pump power to 21 dBm which enhances the loss-gain-ratio. Hence, the maximum fractional time delay could be increased to three. The gain pump power was varied from 14 dBm to 21 dBm. As can be seen from Figure 5.16a, the time delay is reduced if only one or no TFBG is used. This is due to the insufficient or non-blocking of the spurious Stokes-wave.

The highest time delay enhancement of approximately 12 % was obtained by using a WDM filter for the blocking. Due to the insertion loss of the filter, the first and second fiber segments had a length of 2 km and 25 km. This results in smaller equal gains in both SSMF because the Brillouin threshold is much higher for the short first fiber segment. Therefore, a smaller maximum fractional time delay of approximately 2.1 than for the aforementioned measurements

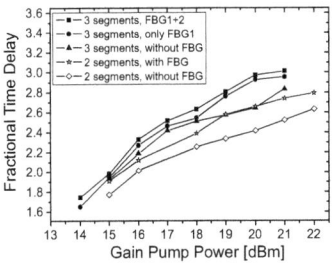

(a) Fractional time delay for blocking with FBG.

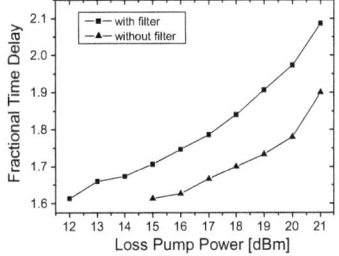

(b) Fractional time delay for blocking with filter.

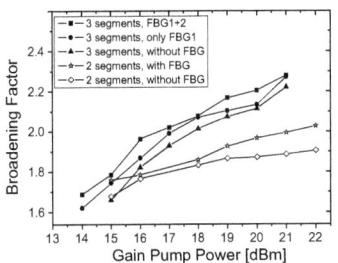

(c) Broadening factor for blocking with FBG.

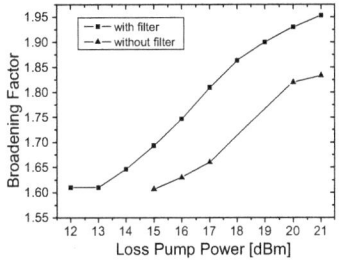

(d) Broadening factor for blocking with filter.

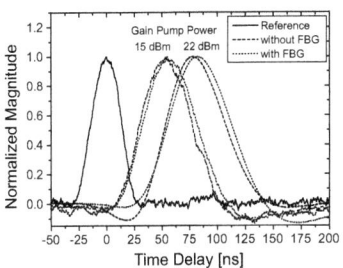

(e) Normalized time functions of selected delayed pulses for blocking with one FBG.

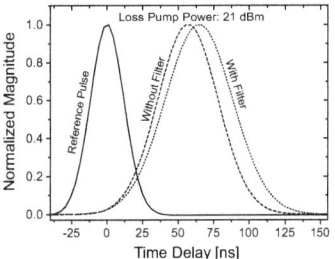

(f) Normalized time functions of selected delayed pulses for blocking with WDM filter.

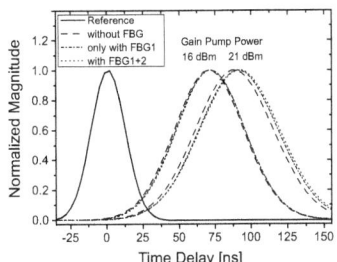

(g) Normalized time functions of selected delayed pulses for blocking with two FBG.

Figure 5.16: Measurement results for time delay enhancement via suppression of backscattered spurious Stokes-waves.

was achieved. This could of course be compensated by applying higher pump powers, as described in Section 3.2. Unforunately, the powers were limited by the used measurement equipment so that a further increase of the time delay was impossible. For the measurements the gain pump power was fixed to 25 dBm and the loss pump power was varied from 12 dBm to 21 dBm. As can be seen in Figure 5.16b, the time delay is increased due to a partial blocking of the spurious Stokes-wave by using the filter centered on the pulse wavelength. Thus, the time delay can be enhanced on average by 10 %. Whereas, the pulse width is only increased on average by 6 % in comparison to the values without filter, as shown in Figure 5.16d.

The normalized time functions of the pulses at selected pump powers in comparison to the reference pulse are shown in Figure 5.16e, Figure 5.16f and Figure 5.16g. From the diagrams it can also be seen that the pulses are further delayed if the TFBG or filter are used between the fiber segments.

Although the used standard filters have an out-band suppression of at least 31 dB, their insufficient low slew rate leads to an attenuation of the spurious Stokes-wave of only 10 dB. To prevent a high attenuation of the loss pump power before entering the second segment, the filters could not be centered on the spurious Stokes-wavelength. Therefore, it can be assumed that a much higher loss-gain-ratio and hence, a much higher enhancement of the time delay is possible if the Stokes-wave is fully blocked by specially designed filters.

5.5 Summary

In this chapter three methods for an enhancement of the maximum achievable time delay were investigated both in theory and experimentally. Since the delay of the pulses is accompanied by an inherent amplification, the delay is limited by the saturation of the Brillouin amplifier. This saturation occurs due to pump depletion if the pump power exceeds the Brillouin threshold. Above the threshold a further increase of the pump power does not lead to a further increase in gain and hence, the time delay. Thus, the time delay is mostly limited to one pulse width ($\Delta T \approx 35$ ns) in the unsaturated regime.

It has been shown that there are two principle methods to mitigate this limitation. At first, the time delay can be decoupled from the amplification gain. Secondly, an increase of the gradient of the spectral gain distribution leads to a higher induced normal dispersion and therefore, to a time delay enhancement. Both methods were realized by a superposition of a Brillouin gain with one or two losses.

A superposition of a narrow Brillouin gain with a broadened loss results only in a decoupling of the time delay from the amplifier gain. By using this method the maximum achievable time delay was improved to approximately 100 ns corresponding to a fractional delay of three. Utilizing a superposition of a Brillouin gain with two losses at its spectral boundaries results additionally in an increase of the gradient of the Brillouin gain distribution. Hence, the presence of the losses leads to a further time delay enhancement. Since this method has several degrees of freedom an optimization of the slow-light performance was shown. The highest time delay

5.5 Summary

obtained with this method was approximately 120 ns which corresponds to a fractional delay of four.

A further time delay enhancement is restricted by the depletion of the loss pumps. If their power exceeds the threshold two spurious backscattered Stoke-waves are created by spontaneous Brillouin scattering which limits the maximum losses. With a third time delay enhancement method it has been shown that also this drawback can be circumvented. Therefore, the depletion of the loss pump powers was reduced by several short fibers with filters in between them to block the spurious Stokes-waves. Hence, a time delay enhancement of on average 10 % was shown with two and three segments of SSMF with TFBG or a WDM filter in between them.

6 Pulse Distortion Reduction

Although the bandwidth and also the time delay limitations can be mitigated, the delayed pulses still experience a pulse width broadening which strongly decreases the effective time delay. Thus, a third line of research investigates an optimization of the slow-light transmission spectrum to minimize the distortions which is the focus of this chapter. At first, the topic is briefly introduced and some related work on pulse distortion reduction is presented. Here, the reasons for the temporal broadening of the delayed pulses are explained and an overview of opportunities to mitigate the broadening effects are given. In the second section of this chapter the background theory on the pulse distortion reduction is described in detail. Therefore, it will be shown that the pulse width of the delayed output pulses can be drastically reduced by a Brillouin bandwidth enhancement and an optimization of a superposition of a Brillouin gain with two losses at its spectral boundaries. For these purposes the experimental results are shown in the third section of this chapter.

6.1 Introduction

As can be seen from the experimental results in the previous chapters, the time delay of the pulses is accompanied by a significant pulse distortion. This distortion primarily manifests itself in a temporal broadening of the pulse width. Although the fractional time delay with SBS-based slow-light can be a few pulse widths, as shown in Chapter 5, the broadened output pulses limit the effective time delay significantly. For example, the large time delay of 120 ns from Subsection 5.3.3 led to a pulse broadening from the initial pulse width of 30 ns to an output pulse width of considerable 85 ns [165]. Hence, the fractional delay of four pulse widths is enormously decreased to 1.4, as can be seen in Figure 6.1. Such a broadening to a triple of

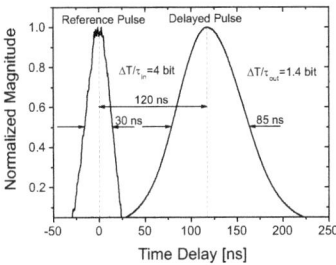

Figure 6.1: Temporal broadening of 120 ns delayed pulse.

the initial pulse width is not suitable for a delay of high data rate signals. Due to occurring intersymbol interferences data losses would be inevitable in practical transmission systems. For these reasons, it is of particular importance to reduce the pulse width broadening in slow-light delays.

The temporal broadening of the delayed pulses has two reasons:

1. From a practical point of view a SBS-based slow-light system can also be seen as a Brillouin amplifier with a limited transmission bandwidth which restricts the pulse bandwidth. Furthermore, as explained in Subsection 3.1.4, the SBS gain has a Lorentzian or Gaussian frequency-distribution. Due to this spectral Brillouin profile the amplification of the pulse is not uniform over the whole frequency range of the Brillouin amplifier. Thus, frequency components which are farther away from the pulse line center are subjected to a lower amplification than the frequency components around the center [214, 215]. Therefore, pulse spectrum is narrowed which leads to a broadening in the time domain after traveling through the SBS-based slow-light system.

2. The second reason for the pulse width broadening originates from the fact that the SBS-based slow-light effect itself is based on a phase change of the pulse and therefore, on an induced pulse dispersion. Due to the different amplification of the distinct frequency components of the pulse the time delay at the center frequency $\Delta T(\omega_0)$ differs from that at the edges of the pulse spectrum. This leads to a distorted pulse. If the pulse distortion is not too strong the GVD-dependent broadening of the pulse can be approximated to the difference between $\Delta T(\omega_0)$ and the time delay at the FWHM bandwidth of the pulse $\Delta T\left(\omega_0 \pm \Delta \omega_P/2\right)$ [214, 216]. Hence, the GVD-dependent broadening factor yields $B_{GVD} = 1 + [\Delta T(\omega_0) - \Delta T\left(\omega_0 \pm \Delta \omega_P/2\right)]/\tau_{in}$.

To circumvent the pulse width broadening due to the delay process the tailoring of the Brillouin spectrum can be used. Basically, the gain-dependent broadening can be reduced by an enhancement of the Brillouin bandwidth [181]. Therefore, the pulse bandwidth will not be restricted by the Brillouin spectrum. Furthermore, a uniform gain profile with a flat top provides an equal amplification of each frequency component of the pulse. Thus, the spectral narrowing of the pulse spectrum is minimized. Additionally, such a broad flat-top SBS gain profile enables a constant time delay over the entire spectral range of the pulse. Therefore, the GVD-dependent pulse broadening is minimized as well. By contrast, the Brillouin bandwidth broadening reduces the achievable time delay. However, for an optimized slow-light performance the time delay should be maximized while the distortions are minimized [217]. To enhance the slow-light delay the gradient of the gain profile has to be increased, as has been shown in Chapter 5. Thus, sharp edges have to be applied to the broad flat-top SBS gain profile [218].

To achieve such a Brillouin spectrum various modulation schemes of the pump spectrum have been proposed. For example, a broad and flat frequency comb was created as pump source by using an external phase modulation resulting in 20 overlapping Brillouin gains [191]. With this approach the broadening factor for pulses with a temporal duration of 5.44 ns was decreased to less than 1.19. Simultaneously a high fractional delay of 2.46 was achieved which corresponds to a high effective time delay of approximately 2.1. A higher flexibility for the improvement of the slow-light performance offers the direct modulation of the injection current of the pump laser with specific shapes of the modulating function. With a pre-designed modulation waveform the pump spectrum can be tailored to generate a Brillouin spectrum with sharp spectral edges leading to stronger gradients in the SBS phase response than those obtained for random pump modulation [219]. Hence, 30 % to 40 % longer delays could be observed for instance for equal pump powers and gain bandwidths. By improving that specific modulation signal, tunable delays up to three pulse widths for 100 ps long input pulses, corresponding to 10 Gbit/s data rates, were proposed while keeping B below 1.8 [220]. This corresponds to an effective time delay of approximately 1.7.

For practical issues the pulse width broadening of the delayed pulses should be minimized or ideally completely avoided. Therefore, such a zero-broadening slow-light would enable the realization of any requested time delay without pulse distortions by cascading such systems. Recently, it was shown that a zero-broadening of the delayed pulses is in fact possible [214, 221–224]. However, this approach is not in the focus of this book and only the reduction of the pulse broadening is described in the following sections.

6.2 Background Theory

According to Equation (3.42), the pulse broadening is decreased for a constant time delay if the Brillouin bandwidth is increased. The broadening factor as a function of the Brillouin bandwidth for a fractional delay $\Delta T_{frac} = 1$ and an input pulse width of 30 ns is shown in Figure 6.2a. As can be seen, B is decreased with Δf_B according to a square root function with a negative exponent. For a maximum practically acceptable broadening of $B = 2$ and a prevention of intersymbol interferences the Brillouin FWHM bandwidth has to be at least 20 MHz. This condition can be fulfilled by all fibers used in the experiments for this work. For $\Delta f_B = 35$ MHz a broadening factor of approximately 1.7 can be expected for a fractional delay of one pulse width. If the natural Brillouin bandwidth is almost doubled to 60 MHz the broadening factor is approximately 1.4. For a further increase of the Brillouin bandwidth the broadening factor approximates a minimum value of 1.3. Additionally, it follows from Equation (3.42) that the maximum gain is limited for a given tolerable broadening factor B by [188, 225]:

$$g \leq \tau_{in}^2 \Delta \omega_B^2 \frac{B^2 - 1}{16 \ln 2}. \tag{6.1}$$

If this gain can really be reached depends on the physical properties of the delay line. The maximum achievable gain is restricted by the saturation of the Brillouin amplifier which in turn depends on the power of the input pulses [225]. Therefore, the highest gain can be achieved for very low input pulse powers. If the input pulse power is not much higher than the noise floor

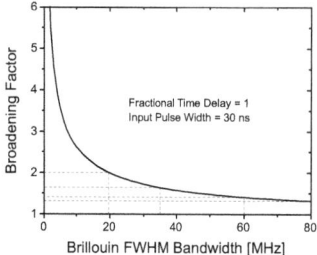
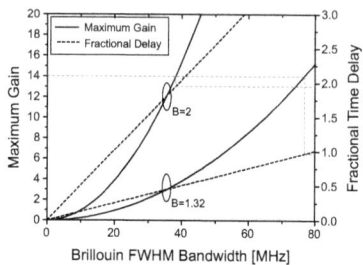

(a) Broadening factor as a function of Brillouin FWHM bandwidth for an input pulse width of 30 ns at a fractional delay of one.

(b) Maximum gain and fractional time delay as a function of the Brillouin FWHM bandwidth for $B = 1.32$ and $B = 2$.

Figure 6.2: Pulse distortion reduction via Brillouin bandwidth broadening.

in the fiber the maximum gain is limited by the threshold of SBS. For a low loss uniform fiber this gain is $g_{th} = 19$ [173]. More realistically is an input pulse power of 1 nW for instance which corresponds to -60 dBm. In this case, the maximum available gain reduces to $g_{-60} \approx 14$ [164]. Since the delay increases with an increasing gain in the small signal regime and it decreases with an increasing gain in the saturation regime [183], for high fractional delays the gain has to be smaller than this maximum value [188].

In Figure 6.2b the maximum gain and the corresponding fractional delay as a function of the Brillouin bandwidth is shown for two different broadening factors. As input pulse width a value of 30 ns was assumed. For a broadening factor of $B = 2$, the maximum fractional delay is $\Delta T_{frac} \approx 1.9$ if the Brillouin bandwidth is 38 MHz. If the broadening factor is reduced to $B = 1.32$ for the same bandwidth the maximum gain is very small. Hence, only a fractional delay of approximately 0.47 can be achieved, as can be seen from the diagram. Thus, for lower broadening factors the bandwidth has to be increased to obtain higher maximum gains and higher fractional time delays. However, the broadening of the Brillouin spectrum leads to a decrease of the fractional delay at the same time. For $B = 1.32$ the maximum fractional delay is only 0.97. This requires a Brillouin FWHM bandwidth of approximately 77 MHz. Therefore, the reduction in time delay due to the bandwidth broadening cannot be compensated by higher pump powers if only a single Brillouin gain is available.

As explained in Section 5.3, the time delay can be drastically enhanced if two losses are inserted at the spectral boundaries of the Brillouin gain. Therefore, the Brillouin spectrum gets sharp edges which increase the gradient of the gain profile and hence, the time delay. According to Equation (5.24), the normalized time delay as a function of the bandwidth ratio k and the loss separation d is shown in Figure 6.3 for equal amounts of the gain and losses ($g_1 = g_2$). Especially if the gain is much broader than the losses ($k \ll 1$), which is for instance the case in high data rate applications, the enhancement of the time delay by the losses can be very high. However, even for $k = 1$ the loss power can compensate the time delay reduction, as can be seen from the diagram.

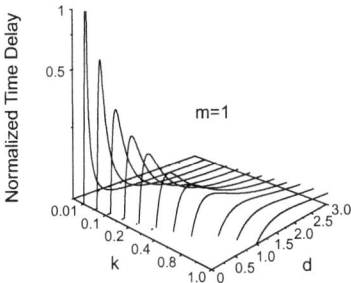

Figure 6.3: Normalized time delay as a function of bandwidth ratio k and the normalized frequency separation of the losses d for a superposition of a Brillouin gain with two losses with a loss-gain-ratio $m = 1$.

For the previous assumptions only the gain-dependent pulse width broadening was considered and the GVD-dependent broadening as well as higher-order dispersion effects were neglected. A reduction of this GVD-dependent broadening can also be achieved by a Brillouin bandwidth enhancement. This results in a broadening of the FWHM bandwidth of the group index distribution. If the pulse is within this bandwidth and if the group index distribution is constant in this region each component of the pulse is equally delayed. Therefore, higher derivatives of the group index, such as the GVD and higher-order dispersions, become zero and a temporal pulse broadening due to GVD can be minimized.

The different opportunities for a Brillouin bandwidth enhancement were described in detail in Chapter 4. These methods can also be used for the pulse distortion reduction in SBS-based slow-light delays. The Brillouin bandwidth broadening and pulse distortion reduction using a polychromatic pump source with discrete spectral lines is shown in Figure 6.4. In the diagrams every case is normalized to its maximum value. With a higher number of pump lines the overall Brillouin spectrum becomes broader. At the same time the frequency-distribution of the group index change is broadened and its top becomes very flat, as can be seen in Figure 6.4c. Hence, the GVD is significantly reduced around the line center of gain spectrum. For a pulse whose spectrum is within this region a low pulse width broadening can be expected. However, by broadening the gain the refractive index slope and therefore, the normal dispersion is decreased, as can be seen in Figure 6.4b. Finally, this leads to a reduction of the time delay. Furthermore, the several pump lines have to be close together to create a uniform Brillouin gain without ripples. Their distance from the line center should not be higher than a multiple of $\gamma/\sqrt{2.5}$. By using for instance six independent pump lines the Brillouin bandwidth is only almost doubled, as can be seen in Figure 6.4a. Hence, for ultra-wide Brillouin bandwidths many pump lines are necessary. This is difficult to realize for practical systems.

In contrast to this method, the use of a broadband pump source, e.g. a directly modulated pump laser, is more efficient, as shown in Chapter 4. By using additionally a superposition of

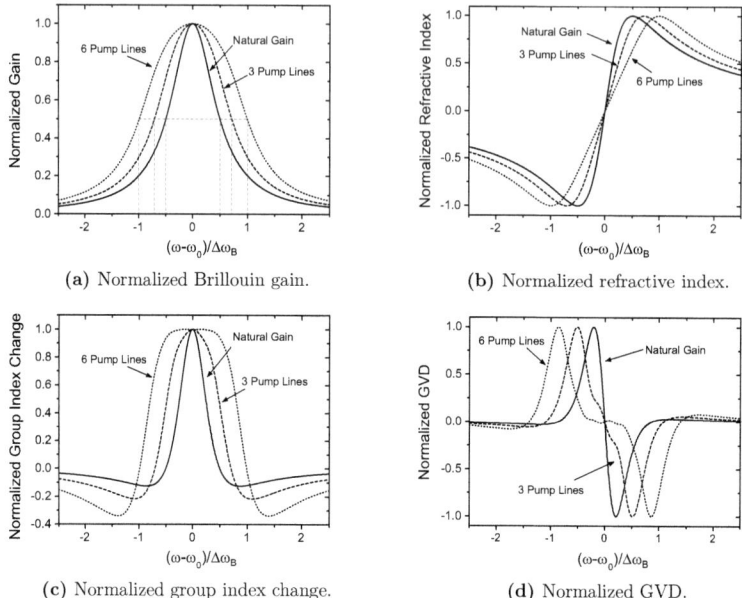

Figure 6.4: Pulse distortion reduction by Brillouin bandwidth broadening via multiple discrete pump lines. Each case is normalized to its own maximum value.

the Brillouin gain with two losses at its spectral boundaries various degrees of freedom for the system optimization are provided, as described in Subsection 5.3.2. This scheme enables ultra-wide bandwidths and is able to enhance the time delay drastically, as described in Chapter 5. Furthermore, the losses can be arranged in a manner that the delay distribution has a very flat top which reduces higher-order dispersion terms [165], as shown in Figure 6.5. Therefore, with this method it is possible to create an optimal Brillouin spectrum for the slow-light delay. By optimizing different parameters, such as the loss-gain-ratio, the loss-gain-bandwidth ratio as well as the frequency separation of the two losses, a broadband flat-top Brillouin profile with sharp edges can be achieved.

Such an optimization is illustrated in Figure 6.6. The diagram shows the frequency-distribution of the normalized group index change and the GVD for different Brillouin spectra. A natural Brillouin gain, a broadened gain and a broadened gain superimposed with two losses at its spectral boundaries with two different frequency separations are compared [226]. For this purpose the group index changes were normalized to the maximum value of the natural gain case. As can be seen from Figure 6.6a, a high group index change and hence, a high time delay in the line center can be achieved with a natural gain. In this case the GVD is very high as well, as can be seen in Figure 6.6b, and the pulses will experience a large temporal broadening. A much higher group index change and hence, a higher effective time delay could be achieved

6.2 Background Theory

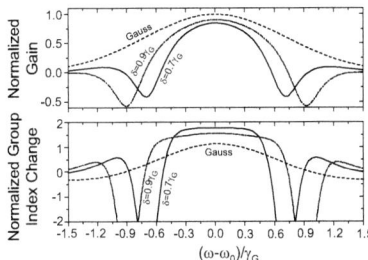

Figure 6.5: Normalized gain and line center group index change for a pure Gaussian-shaped Brillouin spectrum and a Gaussian-shaped Brillouin spectrum superimposed with two losses at its spectral boundaries.

by adding two loss spectra at the spectral boundaries of the normal gain, as explained in Section 5.3. However, this is not shown here because it does not lead to a reduction of the GVD. If the gain is broadened (in this case to $\gamma_G = 3\gamma_2$) on the one hand, the GVD is decreased drastically and it runs linearly with a small slope in the middle of the Brillouin profile. On the other hand, the group index in the line center of the distribution is decreased as well. It is only 38 % of the maximum group index change for the natural gain.

With the additional losses at the spectral boundaries of the broadened gain the time delay can be enhanced. Thereby, the frequency separation of the two losses influences the magnitude of the group index change and the slope of the GVD directly. If the losses are far away from the line center their impact is negligible. In contrast to this, the gain bandwidth is constrained with a smaller separation. By varying δ two optimal operating points were determined according to Subsection 5.3.2. At $\delta = \pm 4\gamma_2$ the highest bandwidth of the group index change can be achieved. In this case, the line center group index change is almost 50 % as for a normal gain. Whereas, the GVD-distribution around the line center shows the same progress as for the broadened gain without losses. Hence, the pulse distortions should be at the minimum

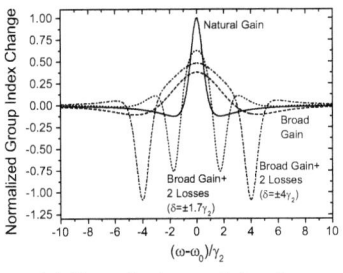

(a) Normalized group index change.

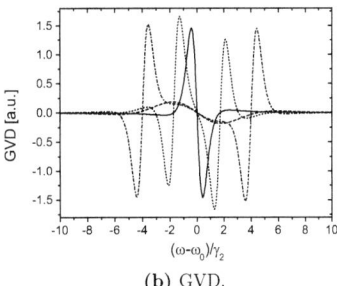

(b) GVD.

Figure 6.6: Normalized group index change and GVD for a natural gain (solid), a broadened gain (dashed, $\gamma_G = 3\gamma_2$) and a broadened gain superimposed with two losses at $\delta = \pm 1.7\gamma_2$ (dotted) and $\delta = \pm 4\gamma_2$ (dashed-dotted).

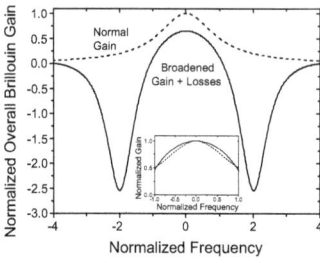
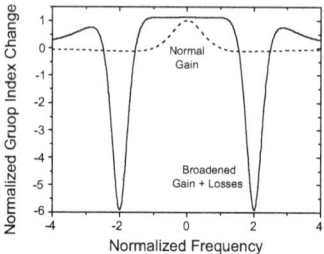

(a) Normalized overall Brillouin gain spectrum. The inset shows a zoom into the gain spectra for a vertically shifted solid line.

(b) Normalized group index change.

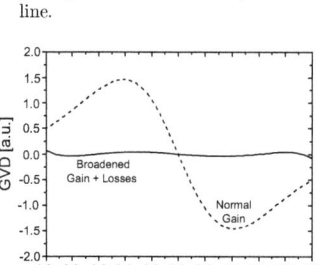
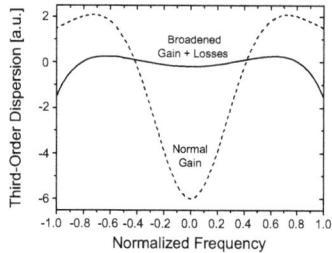

(c) GVD (Derivative of the group index change).

(d) Third-order dispersion (Second derivative of the group index change).

Figure 6.7: Behavior of a superposition of a doubled Brillouin gain with two losses at its spectral boundaries in comparison to a single gain. The parameters are: $m = 3$, $k = 0.25$ and $d = 1$.

for this adjustment. In contrast to this, the highest group index change for this case can be achieved by a loss separation of $\delta = \pm 1.7\gamma_2$. Then, the peak value of group index change is 63 %. This is at the expense of the GVD, which runs with an even steeper slope in the middle of the Brillouin profile. However, the slope is still smoother than for the natural gain case. Hence, for this adjustment the time delay is still high while the pulses experience only a small temporal broadening.

If the Brillouin spectrum is very broad and flat the GVD and the higher-order dispersions could even be reduced to zero within the flat gain region around the line center. If the spectral components of the pulse do not exceed this region the pulse does not experience a temporal broadening due to the GVD. This is shown in Figure 6.7. The diagrams show the behavior of a single Brillouin gain and a doubled Brillouin gain superimposed with two losses for $m = 3$, $k = 0.25$ and $d = 1$. As can be seen from the inset in Figure 6.7a (the superposition case was vertically shifted), the bandwidth of both gains is almost equal but for the superimposed case the spectrum is more flat. Therefore, the group index change around the line center is flat as well, as can be seen in Figure 6.7b. In Figure 6.7c and Figure 6.7d the first and second derivatives of Δn_g are shown. As can be seen from the diagrams, due to the flat group index change the higher-order dispersions are almost zero within the bandwidth region. Hence, both the gain- and GVD-dependent pulse width broadening is minimized as long as the pulse spectrum is within this region.

6.3 Experimental Verification and Results

At first, the pulse distortion reduction is practically investigated using several independent pump lines. Additionally, this method was applied to a cascaded slow-light system consisting of two segments to provide a sufficiently high time delay [196]. Therefore, the pump and slow-light medium setup was adapted as shown in Figure 6.8a. The pulses with a temporal duration of 42 ns were delayed in a two-stage delay line, each consisting of a SSMF with a length of 50 km. Both fibers are pumped separately but by the same pump source. To produce a Brillouin bandwidth broadening by discrete pump lines, the pump LD was externally modulated, as explained in Subsection 4.2.1. With a modulation frequency of 17.5 MHz this leads to three pump lines whose amplitude was adjusted to be equal to one another. Since the two fibers differ slightly in their Brillouin shifts, six independent Brillouin gains overlap if both segments are used. The pump power for the segments and hence, the time delay were controlled by an

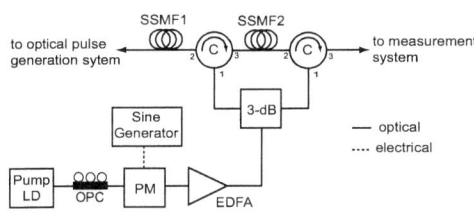

(a) Experimental pump system setup.

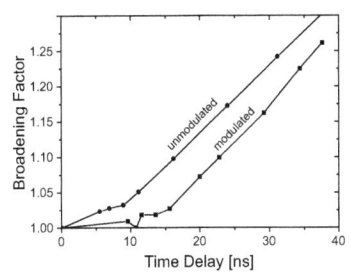

(b) Pulse broadening as a function of the time delay for one segment with a modulated and unmodulated pump source.

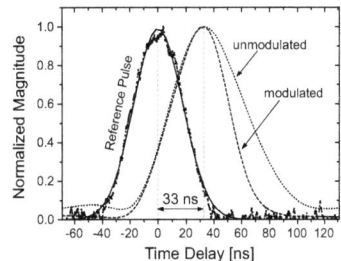

(c) Normalized time functions of output pulses of a two-segment delay line for a modulated and unmodulated pump source.

Figure 6.8: Experimental verification of pulse distortion reduction in cascaded slow-light systems via external pump modulation.

EDFA.

The measured broadening factor as a function of the time delay for only one segment with a modulated and unmodulated pump is shown in Figure 6.8b. In general, the pulse broadening increases with the time delay for both cases. To achieve the same time delays the modulated pump requires a higher pump power. At the same time the resulting pulse broadening is smaller than for an unmodulated pump source, as can be seen from the diagram. The time functions of all pulses for this measurement are shown in Section B.6.

If two segments are used this effect is even more obvious. The output pulses of a two-segment delay line for a time delay of approximately 33 ns are shown in Figure 6.8c. Contrary to the delayed pulses, the reference propagates through the fibers without amplification. Hence, it is rather small and noisy. For the time delay the input pulse was adjusted to a constant power of −12 dBm in the unmodulated case. Both segments were pumped with an optical power of 6.33 dBm. The resultant output pulse had a temporal width of approximately 60 ns, which corresponds to a broadening factor of 1.42. To achieve the same time delay for the modulated case, the amplifier gain has to be increased in the two segments. Hence, the input pulse power was reduced to −31.4 dBm and the pump powers were increased to 11.5 dBm. According to Equation (3.42) the higher gain would result in a higher broadening of the pulse. However, since the Brillouin bandwidth is drastically enhanced at the same time, the resultant pulse broadening is reduced, as can be seen in Figure 6.8c. For a time delay of 33 ns the measured output pulse width was approximately 48 ns which is only 1.14 times that of the input pulse. Hence, the distortion reduction in the two-segment slow-light system is approximately 25 %. By adding more Brillouin gains with additional pump sources this method could be further enhanced.

For the distortion reduction by the superposition of a broadened Brillouin gain with two losses at its spectral boundaries the same setup as described in Subsection 5.3.3 and shown in Figure 5.9 was used. Thereby, pulses with a temporal width of 30 ns were delayed in a SSMF with a length of 50 km. The gain bandwidth was almost doubled to approximately 60 MHz. To achieve a high time delay with the superposition the frequency separation between the two losses was also approximately 60 MHz ($\delta \approx 1.7\gamma_2$ which corresponds to $f_{mod} \approx 30$ MHz). By changing the pump powers of the gain and the losses the broadening factor as a function of the achieved fractional time delay was measured for different cases [188]. The measurement results are shown in Figure 6.9.

The pulse broadening factor as a function of the fractional delay can be seen in Figure 6.9a. For the first case the pulses were delayed within a single natural Brillouin gain without additional losses (square symbols). For this measurement the gain pump power was varied between −3 dBm and 7 dBm. As can be seen from the diagram, the pulses become broader with higher time delay values according to the square root function in Equation (3.42). Thereby, for low fractional delays the measured values deviate from the fitting curve. This can be addressed to uncertainties in the determination of the pulse width because low delays are accompanied

6.3 Experimental Verification and Results

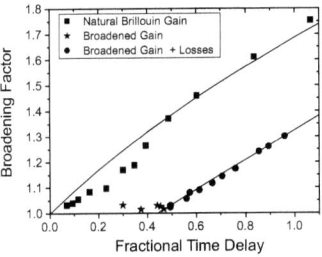

(a) Broadening factor as a function of the fractional time delay for a natural Brillouin gain, a broadened gain and a broadened gain superimposed with two losses at its spectral boundaries.

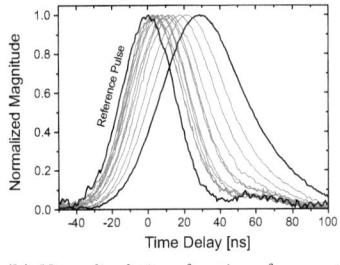

(b) Normalized time functions for a natural Brillouin gain.

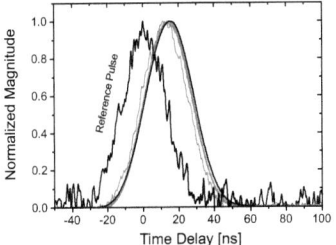

(c) Normalized time functions for a broadened Brillouin gain.

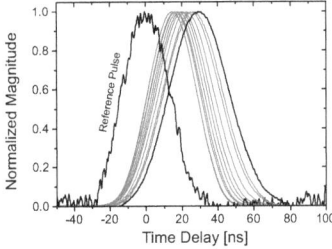

(d) Normalized time functions for a broadened Brillouin gain superimposed with two losses at its spectral boundaries.

Figure 6.9: Measurement results for the distortion reduction via superposition of a Brillouin gain with two losses at its spectral boundaries.

with low pulse powers for the very long fiber. As expected from the theory, for a fractional delay of one, which corresponds to an absolute time delay of 30 ns, the broadening factor was approximately 1.72. The large broadening with an increasing time delay can be clearly seen from the normalized time functions of the delayed pulses in Figure 6.9b.

If the bandwidth of the Brillouin gain is doubled by a direct modulation of the pump laser with a noise signal the broadening factor can be drastically reduced, as shown by means of the star symbols in Figure 6.9a. The output pulses are only slightly broadened which can also be seen in Figure 6.9c. However, as expected from theory, the maximum fractional delay is reduced as well. For a pump power of around 14 dBm it is only around 0.5. If the pump power is increased further the delay line goes into the saturated regime which can be seen by a stagnation or even a reduction of the fractional delay values.

Higher fractional delays can only be achieved by the incorporation of additional losses, as shown by the circle symbols in Figure 6.9a. If the pump power of the gain is kept constant and the loss pump power is increased the time delay increases as well. However, in comparison to the normal gain case an in fact smaller pulse broadening for the same time delays can be achieved, as shown in Figure 6.9d. For a loss pump power of 12 dBm a fractional time delay of one can be obtained again. In this case the broadening factor of only 1.32 is much smaller. This corresponds to a distortion reduction of approximately 25 % for the same fractional delay. If a gain is superimposed with additional losses Equation (3.42) no longer holds. Hence, as can be seen from the fitting curve in Figure 6.9a, the dependence between B and ΔT_{frac} is rather linear for this arrangement.

Although this measurement has only shown the distortion reduction for pulses with a rather low bandwidth, it can be expected that this method is also suitable for high data rate systems. Furthermore, as already explained in Section 5.4, the increase of the losses is restricted by the pump depletion because of arising backscattered spurious Stokes-waves. Therefore, higher fractional time delays at small distortions could be achieved by utilizing a blocking of these waves, as shown in Subsection 5.4.2.

6.4 Summary

Although the time delay can be enhanced to a multiple of the initial pulse width, the effective time delay is significantly reduced by the temporal broadening of the output pulses which is accompanied by the slow-light delay. Therefore, in this chapter this pulse broadening was analyzed and methods were investigated to reduce the distortions. At first, it was shown that the pulse broadening is caused by two reasons: a spectral narrowing due to the Brillouin gain spectrum and SBS-induced GVD. Therefore, a tailoring of the Brillouin spectrum can also reduce the pulse distortions.

In principle, both the gain-dependent and GVD-dependent broadening can be mitigated if a broadened Brillouin gain with a flat top and sharp edges is applied. For the bandwidth enhancement the same methods as described in Chapter 4 can be used. Therefore, a pulse broadening reduction of approximately 25 % was achieved in a cascaded slow-light delay line by using multiple pump lines due to an external modulation of the pump source. An equal amount of distortion reduction with a time delay up to one initial pulse width was shown by a superpostion of a broadened Brillouin gain with two losses at its spectral boundaries.

7 Fast-Light

In the previous chapter the slowing down of optical signal pulses by SBS has been discussed in detail. In recent years the worldwide research focused on this rather than on the acceleration of optical pulses. The reasons for the larger importance of SBS-based slow-light systems are their easier realization and that the practically achievable positive time delays are much higher [227]. Furthermore, potential applications, such as the synchronization of different signal channels, can be accomplished by using only positive time delays just as well. In contrast to this, the advancement of the pulses is more complicated since it is bound to an absorption resonance. Hence, a large negative time delay of the pulses is accompanied by a strong attenuation which makes the detection of the output signals very difficult. Nevertheless, there have also been some investigations on fast-light. In this chapter the principle operation of SBS-based fast-light is briefly explained. Therefore, only one possible fast-light configuration consisting of a superposition of broad Brillouin gain with a narrow Brillouin loss is described theoretically and investigated practically.

7.1 Introduction

As explained in Section 3.2, SBS-based fast-light works in principle similar to slow-light. It has the same behavior and the same conditions are valid, but in the opposite manner. The acceleration of the optical pulses can be realized by using an anti-Stokes Brillouin loss instead of a gain. According to the KKR, as explained in Section 2.2, this leads to an anomalous dispersion and hence, to a decrease of the group index. Thus, the group velocity is increased and can even become higher than the speed of light in a vacuum. However, as described in Subsection 2.1.4, superluminal group velocities do not contradict Einstein's causality because information travel with a different velocity, the signal velocity, whose maximum value can always be equal to c. In the case of SBS-based fast-light the Brillouin gain g and the time delay become negative in Equation (3.40). In the first experiments of slow- and fast-light the advancement of 100 ns wide pulses were approximately -10 ns in a SSMF with a length of 11.8 km and -14.4 ns in just a 2 m fiber showing a large negative group velocity [13, 15]. The tuning of the time delay was applied by controlling the pump power for the Brillouin loss.

In another experiment it was shown that the pulse acceleration can also be facilitated in the absence of an additional pump source by the signal pulse itself [228]. If a signal, whose average power is above the Brillouin threshold, is launched into the fiber the process of spontaneous Brillouin scattering will create a substantial frequency-downshifted Stokes-wave. This sufficiently powerful Stokes-wave depletes at high signal powers and will in turn induce a frequency-upshifted Brillouin loss in the spectral region of the pulses. Associated to this loss is a spectral region of anomalous dispersion in which the pulse will experience an advancement due to fast-light. With this technique pulses with a temporal FWHM width of 45 ns have been advanced by approximately 12 ns [228].

As can be seen from these examples, the achievable advancement time due to fast-light is lower than the time delay through slow-light. The problem is that the advancement inside the loss is accompanied by a strong inherent absorption of the pulse, so that it cannot be detected if the loss pump power increases too much. A mitigation of this limitation is possible by using a technique similar to EIT, but with a reversed characteristic: a narrow absorption window within a broad gain resonance. Thus, the base line of the loss is shifted into a gain region. The result is that the absorption of the pulses is reduced and zero-loss fast-light can be provided.

Such a scheme is realizable through a superposition of multiple Brillouin (gain or loss) spectra as well. For example, if two independent Brillouin gains partially overlap, so that a small dip occurs between them, the group index slope becomes negative in the center of this profile. With this approach, pulses with a FWHM width of 37 ns have been advanced by approximately 4 ns [229]. A similar spectral Brillouin profile can be achieved by cascading two distinct fibers showing slightly different Brillouin shifts [230]. The spectral broadening and the power of the single pump wave defines the size of the absorption dip between both gains. In doing so, a 0.04 ns/mW advancement slope for 25 ns pulses was obtained.

Another opportunity to realize the reverse EIT spectral profile is to superimpose a broad Brillouin gain with a narrow Brillouin loss in its center. The theory and practical results of this method are described in the following sections. By using this technique, a lossless advancement of 50 ns pulses up to approximately 7 ns has been shown elsewhere independently from these investigations [202].

7.2 Background Theory

As described in Section 5.2, the characteristic of the superposition of a broad Brillouin gain with a narrow Brillouin loss is opposite to that of the superposition of a broad Brillouin loss with a narrow gain. However, to describe this method theoretically the complex wave number is equal to Equation (5.1). Hence, the resulting advancement time is equal to Equation (5.6):

$$\Delta T = \frac{1}{2\gamma_1}\left(g_1 - \frac{g_2}{k}\right) = \frac{1}{2\pi\,\Delta f_{B1}}\left(g_1 - \frac{g_2}{k}\right). \tag{7.1}$$

In this case the bandwidth ratio between the loss and the gain is $0 < k < 1$. Therefore, ΔT is negative which corresponds to a pulse advancement. If the gain g_1 is equal to the loss g_2 the pulses are accelerated without any amplitude changes.

The behavior of such a superposition of the broad gain with the narrow loss is shown in

7.2 Background Theory

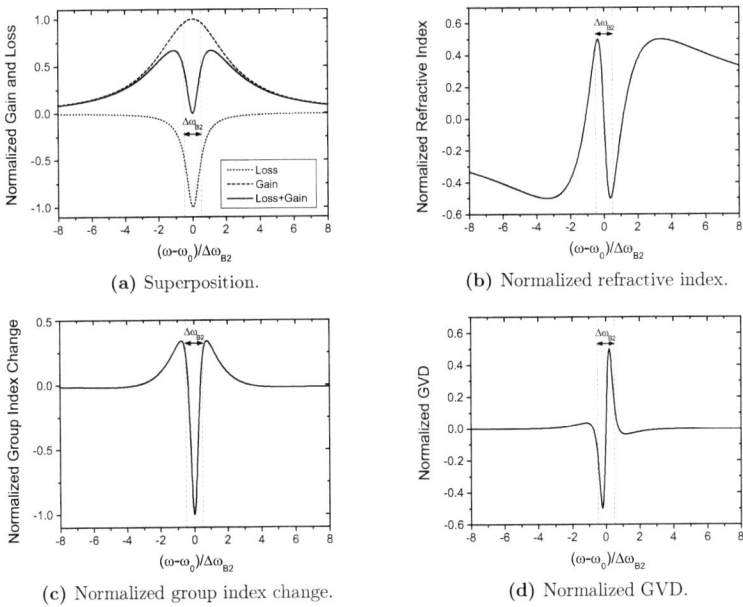

(a) Superposition.

(b) Normalized refractive index.

(c) Normalized group index change.

(d) Normalized GVD.

Figure 7.1: Behavior of a superposition of a narrow loss with a broadened gain.

Figure 7.1. As can be seen in Figure 7.1a, the base line of the Brillouin loss is shifted into a gain region by the broad Brillouin gain. The magnitude of the final Brillouin gain (or absorption) in the line center of the distribution depends on the ratio between the gain and the loss. This means that the advancement time, similar to the slow-light case, is decoupled from the Brillouin amplification and absorption, respectively. Thus, the additional attenuation of the pulses due to Brillouin absorption does not restrict their detection as long as the gain is equal or slightly higher than the loss. Even for a final line center gain value greater than zero the pulses will be advanced while they are amplified. Only if the line center gain value comes into the loss region the additional absorption leads to a strong attenuation of the pulses due to the fast-light effect.

The spectral hole within the gain profile leads to a strong anomalous dispersion within the loss bandwidth, as can be seen in Figure 7.1b. Therefore, the group index change is negative resulting in a decrease of the group index and an increase of the group velocity. According to Equation (7.1), the pulse advancement depends on the line center values of the gain and the loss as well as on their bandwidths. In practice, the time delay can be tuned by controlling the loss pump power if the bandwidths and the gain is fixed. If the loss pump power exceeds the Brillouin threshold also the loss will saturate which limits the maximum achievable negative time delay. However, similarly to the explanations in Section 5.2, the advancement times can be much higher due to the superposition of the loss with a broad gain resulting in a delay-gain-decoupling or in this case in a delay-absorption-decoupling.

In contradiction to the slow-light case, the GVD for the fast-light shows a positive slope within the loss bandwidth, as can be seen in Figure 7.1d. This leads to a reshaping of the pulse as a direct consequence of classical interference between its different frequency components in an anomalous dispersion region [25]. The result of this reshaping can be a narrowing of the pulse in the time domain. Therefore, as will be shown in the next section, the pulse maximum will arrive earlier at the system output while the position of the pulse's leading edge remains at the same position as for the propagation without fast-light. Thus, the trailing edge will arrive earlier as well, which results in the temporal narrowing. Besides this, the pulse reshaping can also lead to additional pulse distortions. For long propagation distances the pulse spectrum distorts dramatically and becomes multi-peaked under certain conditions [227]. Therefore, the pulse shows a ringing in the time domain, as will also be demonstrated in the measurement results in the next section.

7.3 Experimnental Verification and Results

The superposition of a broadened Brillouin gain with a narrow Brillouin loss to produce fast-light was practically investigated with the same setup as described in Subsection 5.2.2 and shown in Figure 5.2. Therefore, either the carrier wavelengths of the two pump waves have to be swapped or the direct modulation with the noise has to be changed to pump LD1. Inside the slow-light medium, in this case a SSMF with a length of 50 km, the overall Brillouin spectrum occurs due to the superposition of the narrow loss within the broadened gain. The final Brillouin spectrum was measured according to the descriptions in Section A.3 and can be seen in Figure 7.2a. Therefore, the Brillouin gain shows a FWHM bandwidth of approximately 180 MHz.

For the measurement of the advancement times of the pulses with a temporal FWHM width of 35 ns the loss pump power was varied by the VOA between 0 mW and 5.5 mW. The negative time delays were determined for different fixed gain pump powers, i.e. 5.5 mW, 8.7 mW, 13.8 mW and 21.9 mW. The results of this measurement have been presented in [20] and are shown in Figure 7.2b. According to Equation (7.1), the negative time delay increases with increasing loss pump powers. Additionally, the negative time delay becomes even higher for the same loss pump power if the gain pump power is increased. On the one hand, the higher gain provides an improved detection of the strongly absorbed pulses. On the other hand, with a higher gain the delay-absorption-decoupling is improved and the saturation is decreased similarly to the slow-light case, as explained in Section 5.2.

As can be seen in Figure 7.2b, for moderate gain pump powers (5.5 mW and 8.7 mW) the time delay as a function of the loss pump power is almost linear and starts with a zero-delay. By increasing the loss pump power the negative time delay also increases up to approximately -19 ns and -22 ns which corresponds to fractional time delays of -0.55 and -0.63. The measurement of higher advancement values was impossible because the large absorption of the pulses

7.3 Experimnental Verification and Results

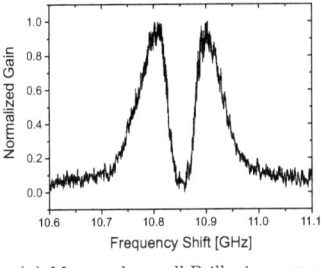

(a) Measured overall Brillouin spectrum.

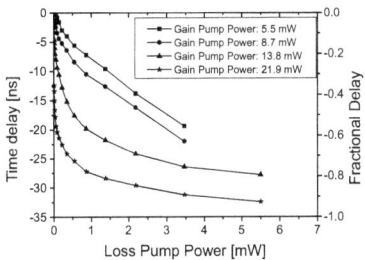

(b) Time delay as a function of the loss pump power.

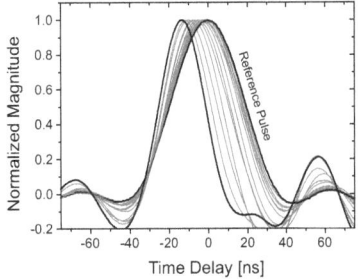

(c) Example of normalized time functions of accelerated pulses in comparison to the reference pulse.

Figure 7.2: Results of fast-light measurements.

prohibited the pulse detection. However, by using higher gain pump powers also higher loss pump powers can be utilized and the negative time delay can be further increased, as can be seen in Figure 7.2b. For gain pump powers of 13.8 mW and 21.9 mW the curves already start with a negative time delay. This is due to the high gain pump power which drives the Brillouin gain into saturation. Thus, the leading edge of the pulse can experience a higher amplification than the trailing edge which results in a pulse advancement [183]. At the same time also the Brillouin loss saturates for high loss pump powers. Therefore, the curves for the negative time delays increase exponentially. The highest achieved pulse advancement was approximately -32.4 ns which corresponds to a fractional time delay of almost minus one.

An example of accelerated light pulses by SBS-based fast-light in comparison to the reference pulse is shown in Figure 7.2c. As can be seen from the diagram, the pulse maximum arrives earlier with increasing loss pump powers which causes the negative time delay. At the same time the leading edges of all pulses occur at the same point in time and therefore, on the one hand the pulse width is decreased due to the pulse reshaping. On the other hand, this reshaping is additionally accompanied by distortions in front of and behind the pulse. These distortions lead to a ringing and forms a small satellite pulse. However, in optical communication systems where the logical 'one' is detected at 50% of the maximum pulse power this might not be crucial. Since the additional pulse is very weak and below 50% of the main pulse, it would only increase the signal noise.

7.4 Summary

The previous investigations on slow-light showed that a high time delay can be realized very easily and efficiently. In contrast to this, the pulse advancement is much more complicated because it is accompanied by a large pulse attenuation. Therefore, only the principle operation of SBS-based fast-light was described and one system was investigated experimentally in this chapter. However, since fast-light works directly opposed to slow-light the same conditions but with an inverse characteristic could be assumed.

Fast-light is achieved inside a Brillouin absorption resonance. Therefore, the pulse advancement is controlled by the pump power of the loss. Due to the absorption of the pulse their detection is very difficult and only small advancement times could be measured. As it was shown, this can be improved by superimposing a narrow Brillouin loss with a broadened Brillouin gain. Thus, the advancement time is decoupled from the absorption resulting in a measured maximum advancement time of one initial pulse width. Furthermore, it could be seen from the measurements that fast-light leads to a reshaping of the pulse in the time domain. On the one hand, this can result in a temporal narrowing of the advanced pulses. On the other hand, this causes additional distortions in the pulse shape.

8 Limitations of SBS-Based Slow- and Fast-Light

In Chapters 4 to 6 it has been shown that the disadvantages of common SBS-based slow-light systems, such as narrow bandwidth, limited time delay and pulse broadening, can be mitigated by improved Brillouin spectra. Nevertheless, from a practical point of view a complete prevention of the system constraints is impossible. Thus, the residual limitations are theoretically investigated in this chapter. After an introduction, where related studies on the limits of classical SBS-based slow-light setups are considered, the practical limitations of both proposed superposition schemes, a narrow gain with a broad loss and a gain with two losses are observed in the second section. This includes the maximum achievable Brillouin amplification gain, the maximum achievable time delay and the achievable storage capacity in such systems. Furthermore, the degrading of the pump power due to FWM is examined.

8.1 Introduction

A key figure of merit of SBS-based slow- and fast-light systems is the achievable (fractional) time delay, more precisely the achievable effective time delay or storage capacity which defines the number of pulse widths that can be delayed or advanced [227]. As it was shown in the previous chapters, the dominant competing and impairing effects are bandwidth (pulse broadening) and amplification or absorption (pump depletion). However, in Chapter 5 it was shown that the time delay can be decoupled from the amplification process resulting in a large enhancement of the time delay. Moreover, in Chapters 4 and 6 it was described that the Brillouin bandwidth can be made extremely wide and that the temporal broadening of the delayed pulses can be strongly reduced. Therefore, high effective time delays and a high storage capacity, respectively, are possible. Due to this fact it could be assumed that the SBS-based slow-light delay is theoretically not bound to any fundamental limit [231].

However, in real systems the bandwidth and the saturation, i.e. pulse broadening and saturation-related maximum time delay, are still relevant. In Chapters 5 and 6 it has been shown that a trade-off between both parameters exists which restricts the slow-light efficiency. The effective time delay depends directly on the output pulse widths. Hence, if the pulses are temporally broadened due to spectral reshaping (narrowing) and GVD the effective time delay is strongly limited.

If the pulse broadening is mainly caused by spectral narrowing and a maximum tolerable pulse width broadening of $B = 2$ is assumed the maximum fractional time delay within a transparency window, such as that created by EIT or CPO, can be predicted by [231, 232]:

$$\Delta T_{frac}^{max} = \frac{3}{2}\Delta f_B \, \tau_{in}. \tag{8.1}$$

Hence, if Equation (8.1) is applied to a Brillouin gain resonance with a FWHM bandwidth of 35 MHz a maximum fractional delay of 1.58 corresponding to an absolute time delay of approximately 47 ns could be achieved for input pulse widths of 30 ns. According to Equation (2.21) the maximum effective time delay would be $\Delta T_{eff}^{max} = 0.5\ \Delta T_{frac}^{max}$. Thus, the 'real' time delay is approximately 23.5 ns which corresponds to a storage capacity of approximately 0.8 bit. However, in theory the bandwidth could be arbitrarily high. Furthermore, the pulse width broadening becomes negligible by ensuring that the pulse spectrum is considerably narrower than the Brillouin spectrum [231]. Thus, according to Equation (8.1), the effective time delay could also become very high. As shown in Chapter 4, the Brillouin bandwidth can be in the order of 25 GHz. However, for higher bandwidths more pump sources are necessary and very high pump powers are required to overcompensate the Brillouin losses of the other pumps. This drives the system inevitably into saturation.

Therefore, the dominant limiting factor is the saturation or the maximum achievable gain. At the same time, a sufficiently high gain is necessary to delay the pulse by a certain amount of pulse widths. If a maximum pulse width broadening of $B = \sqrt{2}$ (mainly caused by spectral narrowing of the pulse spectrum) is assumed the required logarithmic peak gain for a storage of a number of bits N in an optical amplifier with a broad flat-top gain spectrum is [233]:

$$G_{dB} \approx 50\,\sqrt{N^3}. \tag{8.2}$$

Thus, to store 3 bit or 4 bit gains of 260 dB and 400 dB, respectively, would be needed. These values are unrealistic as they require extremely high pump powers. Even if such powers would be available in practice the system would suffer from a large induced ASE noise. Therefore, a maximum storage capacity of only 1 bit would be practically possible if a more realistic gain of 50 dB is assumed.

8.2 Limitations of the Investigated Systems

The previous statements are only valid for classical slow-light systems consisting of a single gain resonance. The opportunity to decouple the time delay from the amplification gain in improved SBS-based slow-light systems was not considered to achieve a high storage capacity. Thus, for example for the delay of three or four pulse widths, as shown in Sections 5.2 and 5.3, amplification gain values of only approximately 16 dB and 28 dB were used. Therefore, the limitations must be determined by another way for the improved Brillouin spectra consisting of a superposition of a narrow gain with a broad loss and a superposition of a gain with two losses at its spectral boundaries. In the following section the theoretical investigation of the

8.2.1 Maximum Gain, Maximum Time Delay and Storage Capacity

In general, the optical pulses are delayed within a SBS gain resonance. Hence, the system works as an amplifier where the amplification gain g is responsible for the strength of the delay. If no counterpropagating pulses are present only the noise in the fiber is amplified. For pump powers above the Brillouin threshold this creates a backscattered Stokes-wave, as was explained in Section 3.1. The amplification of this wave reduces the pump power and hence, the gain due to pump depletion. Thus, the maximum available amplification gain saturates and is restricted by the Brillouin threshold. Since the maximum gain directly determines the maximum time delay and therefore, the storage capacity, this is the main limiting parameter in SBS-based slow-light systems.

For a low loss uniform fiber the maximum gain is the threshold gain $g_{th} = 19$ [173]. If the pump wave is modulated or the gain is superimposed with a loss the peak gain is reduced. However, this can be compensated by an increased gain pump power. Therefore, the threshold gain in the line center remains the same if it is assumed that sufficient pump power is available in practice. Only if the pump depletion is very small the time delay increases with an increasing gain [183]. Thus, the gain g_{td} which is responsible for the time delay should be less than g_{th}. In the case of a single pump wave the maximum time delay is achieved with $g_{td} = g_{th} = 19$. However, whether this gain is really reached depends on the saturation of the pulse signal amplification [164]. As in every amplifier, this saturation in turn depends on the power of the input pulses. Therefore, if pump depletion and the fiber attenuation are ignored the maximum amplification gain g_{max} can be written as [164, 225]:

$$g_{max} = \ln\left\{\frac{\exp\left[(1-\xi)\,g_{th}\right] - \xi}{1-\xi}\right\} \quad \text{with} \quad \xi = \frac{P_s(0)}{P_p(0)}. \tag{8.3}$$

For the relation between the output pulse power $P_s(0)$ and the input pump power $P_p(0)$ the conditions according to Figure 3.3 are valid. Therefore, the maximum output pulse power occurs if the input pulse is amplified by the maximum amplification gain. According to Equation (3.12) and ignoring the fiber attenuation it becomes $P_s(0) = P_s(L)\exp\left(g_{max}\right)$. Furthermore, the g_{max} is achieved if $P_p(0) = P_p^{th}$.

By using the fiber parameters listed in Section A.4 and solving Equation (8.3) numerically the maximum amplification gain as a function of the input pulse power can be calculated. For common SBS-based slow-light systems, which are based on a single Brillouin gain only, the time-delay-gain is equal to g_{max}. Figure 8.1a (solid line) shows g_{td} for a SSMF with a length of 50 km. If the input pulse power is very low the output power is low as well. In contrast to this, the amplification gain $g_{max} \rightarrow g_{th} = 19$. For increasing input pulse powers g_{max} decreases. According to Equation (3.40) the maximum time delay decreases as well, as can be seen in

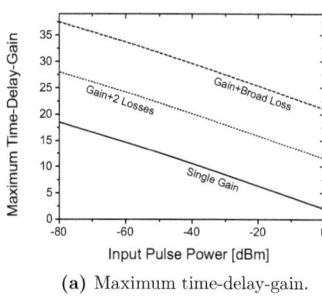 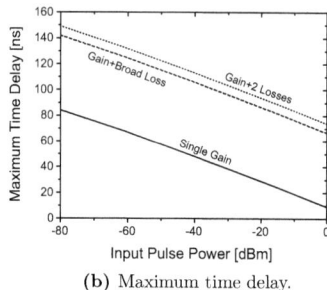

(a) Maximum time-delay-gain. (b) Maximum time delay.

Figure 8.1: Maximum achievable time-delay-gain and time delay as a function of the input pulse power for a SBS-based slow-light system with a single gain, a narrow gain superimposed with a broad loss and a gain superimposed with two losses at its spectral boundaries in a SSMF with a length of 50 km.

Figure 8.1b (for $\Delta f_B = 35\,\text{MHz}$). Hence, maximum time delays between 9 ns and 65 ns can be achieved for input pulse powers of 0 dBm and -60 dBm. If an input pulse width of 30 ns and only a small temporal broadening are assumed the maximum storage capacity is approximately 2 bit at -60 dBm. For a maximum permitted broadening $B = 2$ the maximum storage capacity would be only 1 bit.

However, by using the schemes of superimposed gain and losses $g_{td} > g_{max}$ and hence, a higher storage capacity can be achieved due to the delay-gain-decoupling [234]. If a narrow gain is superimposed with a broad loss the line center amplification gain becomes $g = g_1 - g_2$. Thus, the loss directly reduces the gain. The maximum loss that can be produced is limited by the Brillouin threshold and becomes $g_2 = g_{th} = 19$. Therefore, the maximum amplification gain can be increased by $g_2 = 19$ which leads to an enhanced time-delay-gain of $g_{td} = g_{max} + g_{th}$, as shown in Figure 8.1a (dashed line) for a loss which is three times broader than the gain. The corresponding maximum time delay according Equation (5.5) can be seen in Figure 8.1b. The time delay enhancement is between 1.9 and 7.4 for input pulse powers of -60 dBm and 0 dBm respectively. The maximum time delay of approximately 124 ns for a -60 dBm pulse yields a maximum storage capacity of 4 bit or 2 bit if $B = 2$.

A further improvement of the time delay can be achieved by the superposition of a Brillouin gain with two losses at its spectral boundaries. As described in Section 5.3, for this scheme the line center amplification gain becomes $g = g_1 - g_2/2$. This condition is fulfilled if the gain and losses have the same bandwidth, both losses are equal and the frequency separation of the losses is $2\delta = 2\gamma\sqrt{3}$. The maximum of the losses is again restricted by the Brillouin threshold. Therefore, the maximum time-delay-gain becomes $g_{td} = g_{max} + g_{th}/2$. Although the improvement of the time-delay-gain is not as high as for the aforementioned method, as can be seen in Figure 8.1a, the maximum delay will be further enhanced by the losses according to Equation (5.14). As can be seen in Figure 8.1b, maximum time delays between 74 ns and 132 ns can be expected for input pulse powers between 0 dBm and -60 dBm. In comparison to the single gain case this corresponds to time delay enhancements between 8.2 and 2. Thus, for an input pulse power of −60 dBm the maximum storage capacity is approximately 4.5 bit or 2.25 bit if the output pulse width is doubled.

8.2.2 Pump Power Degradation by Four-Wave Mixing

To achieve the maximum time-delay-gains and maximum time delays as stated above, enough pump power must be available for all pump waves. In principle this is not a problem because SBS in long fibers needs only small pump powers in the range of a few milliwatts. However, for the superposition schemes more than one pump source is used. Therefore, two or three strong pump waves propagate into the same direction. This can result in a parametric interaction between all pump waves due to the third-order nonlinear susceptibility [24]. In consequence, the pump waves generate new co-propagating mixing products via FWM. Since the mixing products counterpropagate to the pulses, on the one hand this will not lead to a decrease of the signal-to-noise ratio or a degrading of the signal quality. On the other hand, the new waves could seriously limit or reduce the maximum pump powers and hence, the maximum time delay [164, 235].

The new mixing products occur at frequencies which are different from the frequencies of the pump waves. According to the energy conservation law the new frequencies are related by [24, Chapter 5]:

$$\omega_{i,j,k} = \omega_i + \omega_j - \omega_k \text{ with } k \neq i,j \text{ and } i,j = 1,2,3, \tag{8.4}$$

where the indices denote the several pump waves. If it is assumed that the pump waves are not depleted, the peak powers of the new generated waves are [24, Chapter 5]:

$$P_{i,j,k} = \left(\frac{D}{3}\beta L_{\textit{eff}}\right)^2 P_{p,i} P_{p,j} P_{p,k} e^{-\alpha L} \eta, \tag{8.5}$$

with $D = 3$ if two pump waves ($i = j \neq k$) and $D = 6$ if three pump waves ($i \neq j \neq k$) are involved in the interaction. $P_{p,i}$, $P_{p,j}$ and $P_{p,k}$ are the input peak powers of the pump waves, β is the nonlinear coefficient and η is the efficiency of the mixing process which depends on the phase matching between the involved waves. By summarizing the power of all mixing products, the strength of the interaction and hence, the degradation of the pump powers can be examined. Therefore, the necessary gain pump power P_{p1} can be determined from the calculated time-delay-gain in Subsection 8.2.1 according to Equation (3.13). The maximum loss pump powers P_{p2} and P_{p3} are equal to the threshold power P_{th} according to Equation (3.4). Since for the pump power calculation a presence of a signal pulse is unnecessary, it can be assumed that $K_B = 1$ in this case. Furthermore, a nonlinear coefficient of $\beta = 1.2\,\text{W}^{-1}\text{km}^{-1}$ [164] and a mixing efficiency of $\eta = 1$ are assumed. Therefore, all pump waves are perfectly matched and co-polarized which means the worst case for the pump power degradation.

If a gain is superimposed with a broad loss two pump waves interact. In this case, only two new waves are generated via FWM. If pump waves 1 and 2 are frequency separated by twice

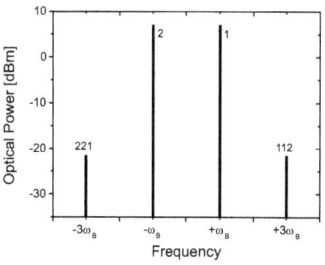
(a) Mixing products for two pump waves.

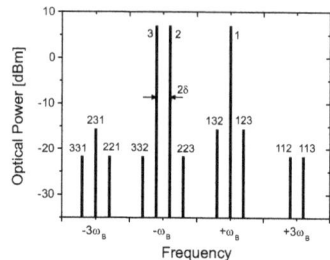
(b) Mixing products for three pump waves.

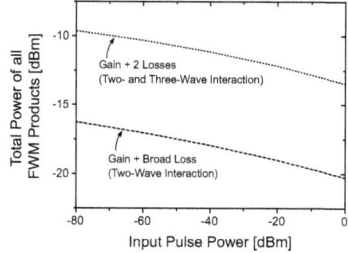
(c) Total power of all generated mixing products for a two- and three-pump wave interaction.

Figure 8.2: Pump power degradation by FWM.

the Brillouin shift, according to Equation (8.4), the mixing products (112 and 221) are up- and downshifted from pump waves 1 and 2 by $2\omega_B$, as can be seen in Figure 8.2a. The dashed curve in Figure 8.2c shows the total power of the two mixing products as a function of the input pulse power for a SSMF with a length of 50 km. Since the maximum gain pump power P_{p1} can be increased with a decreasing input pulse power, the total power of the two-wave interaction increases as well. However, for an input pulse of -60 dBm the maximum mixing power is only approximately -17 dBm. Therefore, it is much smaller than the pump powers used and the pump power reduction by this process is negligible small.

If the gain is superimposed with two losses three pump waves are present and the mixing process also shows a three-wave interaction. Therefore, this results in nine mixing products, as can be seen in Figure 8.2b. Due to the three-wave interaction three of the new waves (123, 132 and 231) are stronger than the others. As shown in Figure 8.2c, the total power of the nine mixing products is higher than for the two mixing products before. For an input pulse power of -60 dBm it is approximately -10.5 dBm. However, according to the discussion in Subsection 8.2.1, the maximum gain pump power P_{p1} can only be half as high as for the superposition with a broad gain. Thus, the increase in total power is not that high and the resultant power is still far below the utilized pump powers, although nine mixing products are created. For these reasons the parametric interaction can be neglected for both cases, the superposition of a gain with a broad loss and also with two losses.

8.3 Summary

In this chapter it was shown that the maximum achievable time-delay-gain, which depends on the saturation of the SBS effect, is the principal limiting factor for the time delay in SBS-based slow-light systems. By decoupling the delay from the Brillouin amplifier gain the time delay performance can be improved. However, the time-delay-gain itself depends on the optical power of the input pulses. Therefore, for high time delays a small input pulse power is necessary. The highest fractional delay and hence, the highest storage capacity is achieved by the superposition of a Brillouin gain with two losses at its spectral boundaries. For a $-60\,\mathrm{dBm}$ input pulse with a width of $30\,\mathrm{ns}$ a maximum storage capacity of approximately $4.5\,\mathrm{bit}$ is possible if the pulse width broadening is negligible. For that, sufficiently high pump powers are required in practice which can result in a nonlinear interaction between the several pump waves. However, as shown in Subsection 8.2.2, a degradation of the pump power due to FWM can be neglected because the Brillouin process is much more effective than the parametric interaction between the pump waves.

9 Conclusions

This book has presented a fundamental investigation of the slow- and fast-light effect on the basis of SBS, in both theory and experiment, toward the application in optical communication and information systems. Therefore, the group velocity of single pulses was altered by exploiting the artificial SBS-induced gain and absorption resonances inside SSMF. Due to the inherent absorption of the pulses in SBS-based fast-light systems slow-light delays can be assigned to be more practicable and effective. In consequence, only one fast-light system was investigated leading to a significant pulse advancement of one pulse width.

The main objective of the work was to find possibilities to overcome the natural limitations of SBS-based slow-light systems. Hence, this work focused primarily on three topics:

1. the enhancement of the Brillouin bandwidth,
2. the enhancement of the SBS-induced time delay and
3. the reduction of the SBS-induced pulse distortions.

For these three topics different novel systems were designed and investigated. The basis of this design is the unique feature of SBS that its spectrum can be tailored by a modulation of the pump spectrum. In doing so, it was shown that all disadvantages of traditional SBS-based slow- and fast-light systems can be successfully overcome.

For the Brillouin bandwidth enhancement two novel methods were investigated. The first approach was an external modulation of the pump source. However, the resulting bandwidth is rather small and a Brillouin bandwidth of 144 MHz was achieved. In comparison to this, a direct modulation of a single pump source with a noise signal is much more effective and results in a maximum achievable bandwidth of considerable 10 GHz. However, it has been shown that the achievable Brillouin bandwidth can be significantly enhanced by using multiple pump sources. If for example two pumps are used the loss of the first can be overcompensated by the gain of the second pump leading theoretically to a Brillouin bandwidth of up to 25 GHz. However, due to the available measurement equipment the experimentally achieved bandwidth was limited to approximately 2 GHz. Nevertheless, the slow-light behavior of such an overcompensated loss has been successfully proven.

Generally, the theory of SBS-based slow-light states that the time delay is proportional to the Brillouin gain and therefore, it is also proportional to the product of the pump power and the fiber length. However, this theory is based on a number of simplifications. Furthermore, many experiments with different fiber lengths have been accomplished in the past. Hence, the practical impact of the fiber length on the achievable amplification gain and time delay was not clear. Preliminary measurements showed that the time delay remains the same in fibers with different lengths (and similar fiber parameters) if the same gain is provided. Therefore, a higher pump power is necessary to achieve the same time delay in shorter fibers. With a single Brillouin gain an equal time delay slope of 0.8 ns/dB was obtained for all SSMF with lengths between 100 m and 50 km. However, as in every traditional SBS-based slow light system, delays of only up to approximately one initial pulse width could be achieved in the unsaturated regime.

To enhance the maximum achievable time delay three novel methods were presented. In comparison to a standard SBS-based slow light system with a single natural gain, these methods include a tailored Brillouin spectrum consting of several Brillouin gains and losses. A first promising approach utilizes a narrow Brillouin gain which was superimposed with a broadened Brillouin loss. Due to the excellent decoupling of the time delay from the amplification gain the time delay was enhanced to 100 ns corresponding to a storage capacity of approximately 3 bit. A further delay enhancement was realized by a superposition of a Brillouin gain with two losses at its spectral boundaries. This method is more efficient because the gradient of the Brillouin gain is additionally increased. Furthermore, this superposition scheme offers several degrees of freedom for the tailoring of the Brillouin spectrum. Hence, an optimization showed that different operating points for a high time delay, a high bandwidth and low pulse distortions can be adjusted. Therefore, a maximum time delay of 120 ns, corresponding to a storage capacity of 4 bit, was obtained which is the highest delay achieved by SBS-based slow-light in one fiber segment so far.

The third time delay enhancement method used a new configuration of the slow-light medium. It consists of several short fibers with filter stages in between them to block the spurious Stokes-waves caused by the additional losses in the multiple pump line systems. This further improves the delay-gain-decoupling and the time delay by approximately 10 %. Unlike proposed cascaded systems, this configuration is less complex and no additional signal noise is induced by the added fiber segments.

To mitigate the SBS-induced temporal pulse broadening two different techniques were investigated. In general, the pulse distortions can be reduced by broadening the Brillouin spectrum. Therefore, in comparison to a standard SBS-based slow-light system a significant distortion reduction of 25 % was demonstrated by an external pump modulation. To preserve a considerable time delay this method was for the first time applied in a cascaded slow-light system. A similar result can be achieved by using a broad Brillouin spectrum with a flat top and sharp edges. This can also be realized by the new Brillouin spectrum tailoring method consisting of a superposition of a broadened gain with two losses at its spectral boundaries. Therefore, besides the time delay enhancement a distortion reduction of 25 % was presented.

Analysis of the system limitations showed that the maximum storage capacity depends on the maximum time-delay-gain which in turn depends on the optical power of the input pulses. Therefore, it was demonstrated that the highest gain can be achieved if the power of the pulses is very small. Furthermore, it was shown that a degrading of the applied pump powers by a parametric interaction between the different pump waves can be neglected. Thus, for the most efficient superposition of a gain with two losses the maximum achievable storage capacity is theoretically limited to 4.5 bit if 30 ns pulses with a power of 1 nW are used and only a small pulse broadening is assumed. In comparison to other slow-light methods this storage capacity is low. Therefore, SBS-based slow-light is less useful for a long-term storage of optical signals.

However, the investigations in this book have shown that SBS is a promising tool for slow- and fast-light systems in optical fibers. Therefore, SBS-based slow-light is a potential method to realize all-optical short-term buffers. In contrast to other techniques, SBS benefits from considerable advantages, such as room temperature operation, small optical power consumption, wavelength independence, simple implementation and compatibility with existing optical communication systems. The unique property to tailor and adapt the Brillouin spectrum gives the systems flexibility and provides opportunities to overcome the natural drawbacks of SBS. Therefore, this work has shown that simple and reliable optically controlled delay lines with storage capacities of a few bits, low distortions and bandwidths capable for high bit rate data signals are possible. Although the slow- and fast-light effect has many potential applications, SBS-based systems may be mainly predestinated for fundamental functions in optical communication and information systems. These especially include systems which require true time delays in terms of a complete phase inversion. Thus, SBS-based slow- and fast-light is useful to realize an accurate synchronization, multiplexing and equalization of multiple data channels. Furthermore, the systems can play an important role in spectroscopy, interferometry as well as in microwave photonics for phased-array devices.

10 Future Work

Although the slow- and fast-light effect on the basis of SBS has been investigated fundamentally there still remains potential for further improvements. Thus, there might be three main issues which have to be considered toward the practical application of SBS-based slow- and fast-light in the future.

A first task could be to further improve the performance of the time delay or storage capacity of the investigated SBS-based slow-light systems. For this, various approaches are conceivable. A first avenue to improve the time delay performance could be to apply other fiber types, such as non-silica fibers which show a high Brillouin gain coefficient g_{Bmax} [4, p. 43]. Therefore, to achieve large time delays the use of very low pump powers while utilizing very short fiber lengths (in the range of a few meters) is possible. For standard silica fibers the Brillouin gain coefficient can exhibit only small maximum values of 5×10^{-11} m/W [158]. In contrast to this, for highly nonlinear bismuth-oxide optical fibers $g_{Bmax} = 6.43 \times 10^{-11}$ m/W [236], tellurite glass fibers show Brillouin gain coefficients up to 2.16×10^{-10} m/W [237–239] and g_{Bmax} of single mode chalcogenite glass fibers can even be around 6×10^{-9} m/W [240–242]. If for all these fiber types the same Brillouin bandwidth, pump power, effective length and effective area are assumed, in comparison to standard silica fibers the achievable time delay would be approximately 1.3 times higher for bismuth-oxide fibers. For tellurite fibers the time delay would be four times higher and for chalcogenite fibers even 120 times. However, due to the high attenuation far exceeding that of silica fibers (up to $0.9\,\text{dB/m}$ [241]), such exotic fibers are only practical for short delay segments (for 5 m of chalcogenite fiber $\alpha = 4.5\,\text{dB}$, for 5 km of SSMF $\alpha = 1\,\text{dB}$) [17].

A further opportunity to improve the slow-light delay performance might be to combine a common SBS-based slow-light system with an external cavity. Therefore, the phase responses of the SBS-based slow-light system and the cavity are added. The increased overall phase slope leads then to an improved delay performance. Such a cavity can be realized by passive resonators, such as Fabry-Perot filters, Bragg structures and coupled ring-resonators for instance [243, 244].

A third method to enhance the storage capacity could be to investigate systems which are able to delay the pulses without distortions. For example, this can be realized by a cascading of systems which provide a high time delay but leading to a pulse broadening and systems which compress the pulse widths and compensates the temporal broadening of the first stage [222–224, 245]. Although such a cascading increases the system complexity, it has the advantage that every time delay without a broadening of the output pulse width could be adjusted.

A second important task would be to investigate the proposed configurations with real data signals. Therefore, the signal spectrum changes with different signal formats. Due to this pattern dependence the Brillouin spectrum has to be adapted to the given format. If data signals with a high bit rate and multiple data channels are used further effects occur inside the slow-light medium which may degrade the performance of the SBS-based slow-light system. For long fibers the chromatic dispersion leads to a strong temporal broadening of the signal pulses. This limits the maximum delayable data rate and has to be carefully considered to avoid intersymbol interferences. Additionally, nonlinear effects, such as self-phase modulation in one channel and cross-phase modulation between multiple channels, are created due to the large temporal variability of the signal intensities. Both effects lead to a spectral broadening of the signal and hence, to a degradation of the system performance. In particular, all these effects determine the capacity of optical transmission systems. Therefore, their influence on the time delay should be determined.

A last step might be the development of a SBS-based slow- and fast-light system as prototype as well as the implementation and investigation of such a system in an existing network. Currently, systems are tested only under laboratory conditions. Therefore, certain engineering issues have to be resolved. For example, a stabilization of the pump wavelength, especially if multiple pump waves are used, is required. In general, the carrier wavelengths of each pump laser fluctuate due to temperature changes. Therefore, there is a relative fluctuation between every pump wavelength resulting in changes of the final Brillouin spectrum. An opportunity to stabilize the system has already been proposed [246]. The method contains the generation of all pump components with only one laser source. Hence, all pump waves experience the same wavelength fluctuation which means that they are stable among each other.

Finally, the engineering system integration plays an important role for the practical application in real systems. Therefore, all adjustments of the system, such as the time delay, should be automated. This includes for example an automatic tracing of the pump wavelengths if the carrier wavelength of the signal is changed as well as an automatic pump power control for a corresponding time delay. On this basis the slow-light system could be investigated under real application conditions. This might open the door for further novel, innovative and promising applications. For example, the impact of reduced group velocity on the interaction of nonlinear processes and on spectroscopy or sensing features is widely unknown until now.

References

[1] C. Schiller. (2007, Nov.) Motion mountain - The Adventure of Physics. 21st ed. [Online]. Available: http://www.motionmountain.net [Accessed: Nov. 7, 2009].

[2] A. Sommerfeld, *Vorlesungen über theoretische Physik Band IV Optik*, 2nd ed. Leipzig: Akademische Verlagsgesellschaft, 1959.

[3] Lord Rayleigh, "On Progressive Waves," *Proc. London Math. Soc.*, vol. 9, no. 1, pp. 21–26, 1877.

[4] J. B. Khurgin and R. S. Tucker, Eds., *Slow Light: Science and Applications (Optical Science and Engineering)*, 1st ed. CRC Press, 2008.

[5] P. W. Milonni, *Fast Light, Slow Light and Left-Handed Light (Series in Optics and Optoelectronics)*. New York: Taylor & Francis Group, 2005.

[6] A. Sommerfeld, "Über die Fortpflanzung des Lichtes in dispergierenden Medien," *Annalen der Physik*, vol. 349, no. 10, pp. 177–202, 1914.

[7] L. Brillouin, *Wave Propagation And Group Velocity*. New York: Academic Press, 1960.

[8] R. S. Tucker, J. Baliga, R. Ayre, K. Hinton, and W. V. Sorin, "Energy consumption in IP networks," in *34th European Conference on Optical Communication (ECOC) 2008*. Brussels, Belgium: IEEE, Sept. 2008, paper Tu.3.A.2.

[9] J. Baliga, R. Ayre, K. Hinton, W. Sorin, and R. Tucker, "Energy consumption in optical IP networks," *J. Lightw. Technol.*, vol. 27, no. 13, pp. 2391–2403, 2009.

[10] G.-K. Chang, K.-I. Sato, and D. K. Hunter, "Guest editorial optical networks (special issue)," *J. Lightw. Technol.*, vol. 18, no. 12, pp. 1603–1605, 2000.

[11] N. Nagatsu, "Photonic network design issues and applications to the IP backbone," *J. Lightw. Technol.*, vol. 18, no. 12, pp. 2010–2018, 2000.

[12] R. Won, "Slow light now and then (Interview with Robert Boyd)," *Nature Photonics*, vol. 2, no. 8, pp. 454–455, 2008.

[13] K. Y. Song, M. González-Herráez, and L. Thévenaz, "Observation of pulse delaying and advancement in optical fibers using stimulated Brillouin scattering," *Opt. Express*, vol. 13, no. 1, pp. 82–88, 2005.

[14] Y. Okawachi, M. S. Bigelow, J. E. Sharping, Z. M. Zhu, A. Schweinsberg, D. J. Gauthier, R. W. Boyd, and A. L. Gaeta, "Tunable all-optical delays via Brillouin slow light in an optical fiber," *Phys. Rev. Lett.*, vol. 94, no. 15, p. 153902, 2005.

[15] M. González-Herráez, K. Y. Song, and L. Thévenaz, "Optically controlled slow and fast light in optical fibers using stimulated Brillouin scattering," *Appl. Phys. Lett.*, vol. 87, no. 8, p. 081113, 2005.

[16] E. Cabrera-Granado and D. J. Gauthier, "Recent advancements in SBS slow light," *Opt. Pura Apl.*, vol. 41, no. 4, pp. 313–323, 2008.

[17] L. Thévenaz, "Slow and fast light in optical fibres," *Nature Photonics*, vol. 2, no. 8, pp. 474–481, 2008.

[18] R. W. Boyd, *Nonlinear Optics*, 2nd ed.   Academic Press, 2003.

[19] A. Wiatrek, "Untersuchung des Laufzeitverhaltens optischer Signale in Abhängigkeit des Gruppenbrechungsindexes in einer Glasfaser," Project IV Thesis, Hochschule für Telekommunikation Leipzig (FH), Leipzig, Germany, 2007.

[20] K.-U. Lauterbach, T. Schneider, R. Henker, M. Junker, M. J. Ammann, and A. T. Schwarzbacher, "Investigation of fast light in long optical fibers based on stimulated Brillouin scattering," in *Conference on Lasers and Electro-Optics and Quantum Electronics and Laser Science (CLEO/QELS) 2007*.   Baltimore, MD, USA: OSA, May 2007, paper JWA48.

[21] R. L. Smith, "The velocities of light," *Am. J. Phys.*, vol. 38, no. 8, pp. 978–984, 1970.

[22] S. C. Bloch, "Eighth velocity of light," *Am. J. Phys.*, vol. 45, no. 6, pp. 538–549, 1977.

[23] M. S. Bigelow, "Ultra-slow and superluminal light propagation in solids at room temperature," Ph.D. dissertation, University of Rochester, Rochester, USA, 2004.

[24] T. Schneider, *Nonlinear Optics in Telecommunications (Advanced Texts in Physics)*. Berlin, Heidelberg: Springer-Verlag, 2004.

[25] J. L. Wang, A. Kuzmich, and A. Dogariu, "Gain-assisted superluminal light propagation," *Nature*, vol. 406, pp. 277–279, 2000.

[26] L. V. Hau, S. E. Harris, Z. Dutton, and C. H. Behroozi, "Light speed reduction to 17 metres per second in an ultracold atomic gas," *Nature*, vol. 397, pp. 594–598, 1999.

[27] C. Liu, Z. Dutton, C. H. Behroozi, and L. V. Hau, "Observation of coherent optical information storage in an atomic medium using halted light pulses," *Nature*, vol. 409, pp. 490–493, 2001.

[28] Z. Zhu, D. J. Gauthier, and R. W. Boyd, "Stored light in an optical fiber via stimulated Brillouin scattering," *Science*, vol. 318, pp. 1748–1750, 2007.

[29] R. Y. Chiao and A. M. Steinberg, "Tunneling times and superluminality," in *Progress in Optics*, E. Wolf, Ed. Amsterdam: Elsevier, 1997, vol. 37, ch. VI, pp. 345–405.

[30] M. D. Stenner, D. J. Gauthier, and M. A. Neifeld, "The speed of information in a 'fast-light' optical medium," *Nature*, vol. 425, pp. 695–698, 2003.

[31] R. W. Boyd and D. J. Gauthier, "'Slow' and 'fast' light," in *Progress in Optics*, E. Wolf, Ed. Amsterdam: Elsevier, 2002, vol. 43, ch. 6, pp. 497–530.

[32] M. D. Stenner, "Measurement of the information velocity in fast- and slow-light optical pulse propagation," Ph.D. dissertation, Duke University, Durham, USA, 2004.

[33] D. J. Gauthier and R. W. Boyd, "Fast light, slow light and optical precursors: What does it all mean?" *Photonics Spectra*, pp. 82–90, Jan. 2007.

[34] J. Jahns, *Photonik: Grundlagen, Komponenten und Systeme*. München, Wien: Oldenbourg Wissenschaftsverlag, 2001.

[35] R. d. L. Kronig and H. A. Kramers, "Zur Theorie der Absorption und Dispersion in den Röntgenspektren," *Zeitschrift für Physik*, vol. 48, no. 3, pp. 174–179, 1928.

[36] B. E. Saleh and M. C. Teich, *Fundamentals of Photonics (Wiley Series in Pure and Applied Optics)*, 2nd ed. Hoboken: Wiley & Sons, 2007.

[37] V. Lucarini, K.-E. Peiponen, J. J. Saarinen, and E. M. Vartiainen, *Kramers-Kronig Relations in Optical Materials Research (Springer Series in Optical Sciences)*, 1st ed. Berlin, Heidelberg, New York: Springer-Verlag, 2005.

[38] C. Kittel, *Einführung in die Festkörperphysik*, 9th ed. München, Wien: Oldenbourg, 1991.

[39] G. A. Reider, *Photonik: Eine Einführung in die Grundlagen*, 2nd ed. Wien, New York: Springer-Verlag, 2002.

[40] R. Tucker, P.-C. Ku, and C. Chang-Hasnain, "Slow-light optical buffers: capabilities and fundamental limitations," *J. Lightw. Technol.*, vol. 23, no. 12, pp. 4046–4066, 2005.

[41] C. Monat, B. Corcoran, M. Ebnali-Heidari, C. Grillet, B. J. Eggleton, T. P. White, L. O'Faolain, and T. F. Krauss, "Slow light enhancement of nonlinear effects in silicon engineered photonic crystal waveguides," *Opt. Express*, vol. 17, no. 4, pp. 2944–2953, 2009.

[42] M. D. Settle, R. J. P. Engelen, M. Salib, A. Michaeli, L. Kuipers, and T. F. Krauss, "Flatband slow light in photonic crystals featuring spatial pulse compression and terahertz bandwidth," *Opt. Express*, vol. 15, no. 1, pp. 219–226, 2007.

[43] W. Frohberg, H. Kolloschie, and H. Löffler, Eds., *Taschenbuch der Nachrichtentechnik*, 1st ed. München: Fachbuchverlag Leipzig im Carl Hanser Verlag, 2008.

[44] G. Gehring, R. W. Boyd, A. L. Gaeta, D. J. Gauthier, and A. E. Willner, "Fiber-based slow-light technologies," *J. Lightw. Technol.*, vol. 26, no. 23, pp. 3752–3762, 2008.

[45] J. Spring and R. Tucker, "Photonic 2*2 packet switch with input buffers," *Electr. Lett.*, vol. 29, no. 3, pp. 284–285, 1993.

[46] J. B. Khurgin, "Optical buffers based on slow light in electromagnetically induced transparent media and coupled resonator structures: comparative analysis," *J. Opt. Soc. Am. B*, vol. 22, no. 5, pp. 1062–1074, 2005.

[47] Q. Xu, P. Dong, and M. Lipson, "Breaking the delay-bandwidth limit in a photonic structure," *Nature Physics*, vol. 3, no. 6, pp. 406–410, 2007.

[48] F. Xia, L. Sekaric, and Y. Vlasov, "Ultracompact optical buffers on a silicon chip," *Nature Photonics*, vol. 1, no. 1, pp. 65–71, 2007.

[49] A. Melloni, F. Morichetti, C. Ferrari, and M. Martinelli, "Continuously tunable 1 byte delay in coupled-resonator optical waveguides," *Opt. Lett.*, vol. 33, no. 20, pp. 2389–2391, 2008.

[50] C. Chang-Hasnain, P.-C. Ku, J. Kim, and S.-L. Chuang, "Variable optical buffer using slow light in semiconductor nanostructures," *Proceedings of the IEEE*, vol. 91, no. 11, pp. 1884–1897, 2003.

[51] P.-C. Ku, F. Sedgwick, C. J. Chang-Hasnain, P. Palinginis, T. Li, H. Wang, S.-W. Chang, and S.-L. Chuang, "Slow light in semiconductor quantum wells," *Opt. Lett.*, vol. 29, no. 19, pp. 2291–2293, 2004.

[52] C. J. Chang-Hasnain and S. L. Chuang, "Slow and fast light in semiconductor quantum-well and quantum-dot devices," *J. Lightw. Technol.*, vol. 24, no. 12, pp. 4642–4654, 2006.

[53] M. van der Poel, J. Mørk, and J. Hvam, "Controllable delay of ultrashort pulses in a quantum dot optical amplifier," *Opt. Express*, vol. 13, no. 20, pp. 8032–8037, 2005.

[54] H. Gersen, T. J. Karle, R. J. P. Engelen, W. Bogaerts, J. P. Korterik, N. F. van Hulst, T. F. Krauss, and L. Kuipers, "Real-space observation of ultraslow light in photonic crystal waveguides," *Phys. Rev. Lett.*, vol. 94, no. 7, p. 073903, 2005.

[55] J. Li, T. P. White, L. O'Faolain, A. Gomez-Iglesias, and T. F. Krauss, "Systematic design of flat band slow light in photonic crystal waveguides," *Opt. Express*, vol. 16, no. 9, pp. 6227–6232, 2008.

[56] T. Baba, "Slow light in photonic crystals," *Nature Photonics*, vol. 2, no. 8, pp. 465–473, 2008.

[57] A. Schweinsberg, N. N. Lepeshkin, M. S. Bigelow, R. W. Boyd, and S. Jarabo, "Observation of superluminal and slow light propagation in erbium-doped optical fiber," *Europhys. Lett.*, vol. 73, no. 2, pp. 218–224, 2006.

[58] J. M. Ezquerro, S. Melle, O. G. Calderón, F. Carreño, and M. A. Antón, "Fractional advancement enhancement in erbium-doped fiber amplifiers by bi-directional pumping," in *Slow and Fast Light Conference 2008*. Boston, MA, USA: OSA, July 2008, paper JMB16.

[59] A. Uskov, F. Sedgwick, and C. Chang-Hasnain, "Delay limit of slow light in semiconductor optical amplifiers," *IEEE Photon. Technol. Lett.*, vol. 18, no. 6, pp. 731–733, 2006.

[60] P. K. Kondratko, H. Su, and S.-L. Chuang, "Fast light using multiple cascaded quantum-well semiconductor optical amplifiers," in *Conference on Lasers and Electro-Optics and Quantum Electronics and Laser Science (CLEO/QELS) 2007*. Baltimore, MD, USA: OSA, May 2007, paper JWA41.

[61] F. Öhman, S. Sales, Y. Chen, E. Granell, and J. Mørk, "Large microwave phase shift and small distortion in an integrated waveguide device," in *Slow and Fast Light Conference 2007*. Slat Lake City, UT, USA: OSA, July 2007, paper STuA6.

[62] Y. Zhang, X. Zhang, X. Huang, and C. Cheng, "Experimental investigation on slow light via four-wave mixing in semiconductor optical amplifier," *Frontiers of Optoelectronics in China*, vol. 2, no. 3, pp. 259–263, 2009.

[63] B. Pesala, F. Sedgwick, A. Uskov, and C. Chang-Hasnain, "Ultrahigh-bandwidth electrically tunable fast and slow light in semiconductor optical amplifiers," *J. Opt. Soc. Am. B*, vol. 25, no. 12, pp. C46–C54, 2008.

[64] W. Xue, S. Sales, J. Capmany, and J. Mørk, "Experimental demonstration of 360° tunable rf phase shift using slow and fast light effects," in *Slow and Fast Light Conference 2009*. Honolulu, HI, USA: OSA, July 2009, paper SMB6.

[65] M. M. Kash, V. A. Sautenkov, A. S. Zibrov, L. Hollberg, G. R. Welch, M. D. Lukin, Y. Rostovtsev, E. S. Fry, and M. O. Scully, "Ultraslow group velocity and enhanced nonlinear optical effects in a coherently driven hot atomic gas," *Phys. Rev. Lett.*, vol. 82, no. 26, pp. 5229–5232, 1999.

[66] A. V. Turukhin, V. S. Sudarshanam, M. S. Shahriar, J. A. Musser, B. S. Ham, and P. R. Hemmer, "Observation of ultraslow and stored light pulses in a solid," *Phys. Rev. Lett.*, vol. 88, no. 2, p. 023602, 2001.

[67] M. S. Bigelow, N. N. Lepeshkin, and R. W. Boyd, "Observation of ultraslow light propagation in a ruby crystal at room temperature," *Phys. Rev. Lett.*, vol. 90, no. 11, p. 113903, 2003.

[68] M. S. Bigelow, N. N. Lepeshkin, and R. W. Boyd, "Superluminal and slow light propagation in a room-temperature solid," *Science*, vol. 301, no. 5630, pp. 200–202, 2003.

[69] G. M. Gehring, A. Schweinsberg, C. Barsi, N. Kostinski, and R. W. Boyd, "Observation of backward pulse propagation through a medium with a negative group velocity," *Science*, vol. 312, no. 5775, pp. 895–897, 2006.

[70] F. Öhman, K. Yvind, and J. Mørk, "Voltage-controlled slow light in an integrated semiconductor structure with net gain," *Opt. Express*, vol. 14, no. 21, pp. 9955–9962, 2006.

[71] J. Sharping, Y. Okawachi, and A. Gaeta, "Wide bandwidth slow light using a Raman fiber amplifier," *Opt. Express*, vol. 13, no. 16, pp. 6092–6098, 2005.

[72] D. Dahan and G. Eisenstein, "Tunable all optical delay via slow and fast light propagation in a Raman assisted fiber optical parametric amplifier: a route to all optical buffering," *Opt. Express*, vol. 13, no. 16, pp. 6234–6249, 2005.

[73] J. Sharping, Y. Okawachi, J. van Howe, C. Xu, Y. Wang, A. Willner, and A. Gaeta, "All-optical, wavelength and bandwidth preserving, pulse delay based on parametric wavelength conversion and dispersion," *Opt. Express*, vol. 13, no. 20, pp. 7872–7877, 2005.

[74] L. Christen, O. F. Yilmaz, S. Nuccio, X. Wu, I. Fazal, A. E. Willner, C. Langrock, and M. M. Fejer, "Tunable 105 ns optical delay for 80 Gb/s RZ-DQPSK, 40 Gb/s RZ-DPSK, and 40 Gb/s RZ-OOK signals using wavelength conversion and chromatic dispersion," *Opt. Lett.*, vol. 34, no. 4, pp. 542–544, 2009.

[75] O. Yilmaz, S. Nuccio, X. Wu, and A. Willner, "10-packet-depth, 40 Gb/s optical buffer with a <0.5 ns reconfiguration time using 116 ns, continuously tunable conversion/dispersion delays," in *Conference on Optical Fiber Communication and the National Fiber Optic Engineers Conference (OFC/NFOEC) 2009*. San Diego, CA, USA: OSA, March 2009, paper PDPC7.

[76] N. Alic, E. Myslivets, S. Moro, B. Kuo, R. Jopson, C. McKinstrie, and S. Radic, "1.83-μs wavelength-transparent all-optical delay," in *Conference on Optical Fiber Communication and the National Fiber Optic Engineers Conference (OFC/NFOEC) 2009*. San Diego, CA, USA: OSA, March 2009, paper PDPA1.

[77] E. Myslivets, N. Alic, S. Moro, B. P. P. Kuo, R. M. Jopson, C. J. McKinstrie, M. Karlsson, and S. Radic, "1.56-µs continuously tunable parametric delay line for a 40-Gb/s signal," *Opt. Express*, vol. 17, no. 14, pp. 11 958–11 964, 2009.

[78] S. R. Nuccio, O. F. Yilmaz, S. Khaleghi, X. Wu, L. Christen, I. Fazal, and A. E. Willner, "Tunable 503 ns optical delay of 40 Gbit/s RZ-OOK and RZ-DPSK using a wavelength scheme for phase conjugation to reduce residual dispersion and increase delay," *Opt. Lett.*, vol. 34, no. 12, pp. 1903–1905, 2009.

[79] Y. Dai, X. Chen, Y. Okawachi, A. C. Turner-Foster, M. A. Foster, M. Lipson, A. L. Gaeta, and C. Xu, "1 µs tunable delay using parametric mixing and optical phase conjugation in Si waveguides," *Opt. Express*, vol. 17, no. 9, pp. 7004–7010, 2009.

[80] N. Alic, C. J. McKinstrie, S. Namiki, and S. Radic, "1 µs tunable delay using para-metric mixing and optical phase conjugation in Si waveguides: comment," *Opt. Express*, vol. 17, no. 18, pp. 16 027–16 028, 2009.

[81] Y. Dai, X. Chen, Y. Okawachi, A. C. Turner-Foster, M. A. Foster, M. Lipson, A. L. Gaeta, and C. Xu, "1 µs tunable delay using parametric mixing and optical phase conjugation in Si waveguides: reply," *Opt. Express*, vol. 17, no. 18, pp. 16 029–16 031, 2009.

[82] S. Preußler, K. Jamshidi, A. Wiatrek, R. Henker, C.-A. Bunge, and T. Schneider, "Quasi-light-storage based on time-frequency coherence," *Opt. Express*, vol. 17, no. 18, pp. 15 790–15 798, 2009, – announced as research highlight in *Nature Photonics*, vol. 3, no. 10, p. 555, Oct. 2009.

[83] M. Rasras, C. Madsen, M. Cappuzzo, E. Chen, L. Gomez, E. Laskowski, A. Griffin, A. Wong-Foy, A. Gasparyan, A. Kasper, J. L. Grange, and S. Patel, "Integrated resonance-enhanced variable optical delay lines," *IEEE Photon. Technol. Lett.*, vol. 17, no. 4, pp. 834–836, 2005.

[84] D. J. Gauthier, A. L. Gaeta, and R. W. Boyd, "Slow light: From basics to future prospects," *Photonics Spectra*, pp. 44–50, March 2006.

[85] M. Fleischhauer, A. Imamoğlu, and J. P. Marangos, "Electromagnetically induced transparency: Optics in coherent media," *Rev. Mod. Phys.*, vol. 77, no. 2, pp. 633–673, 2005.

[86] S. E. Harris, J. E. Field, and A. Imamoğlu, "Nonlinear optical processes using electromagnetically induced transparency," *Phys. Rev. Lett.*, vol. 64, no. 10, pp. 1107–1110, 1990.

[87] S. E. Harris, J. E. Field, and A. Kasapi, "Dispersive properties of electromagnetically induced transparency," *Phys. Rev. A*, vol. 46, no. 1, pp. R29–R32, 1992.

[88] N. S. Ginsberg, S. R. Garner, and L. V. Hau, "Coherent control of optical information with matter wave dynamics," *Nature*, vol. 445, no. 7128, pp. 623–626, 2007.

[89] D. F. Phillips, A. Fleischhauer, A. Mair, R. L. Walsworth, and M. D. Lukin, "Storage of light in atomic vapor," *Phys. Rev. Lett.*, vol. 86, no. 5, pp. 783–786, 2001.

[90] J. J. Longdell, E. Fraval, M. J. Sellars, and N. B. Manson, "Stopped light with storage times greater than one second using electromagnetically induced transparency in a solid," *Phys. Rev. Lett.*, vol. 95, no. 6, p. 063601, 2005.

[91] J. Mørk, R. Kjær, M. van der Poel, and K. Yvind, "Slow light in a semiconductor waveguide at gigahertz frequencies," *Opt. Express*, vol. 13, no. 20, pp. 8136–8145, 2005.

[92] G. Lenz, B. Eggleton, C. Madsen, and R. Slusher, "Optical delay lines based on optical filters," *IEEE J. Quantum Electron.*, vol. 37, no. 4, pp. 525–532, 2001.

[93] E. Choi, J. Na, S. Ryu, G. Mudhana, and B. Lee, "All-fiber variable optical delay line for applications in optical coherence tomography: feasibility study for a novel delay line," *Opt. Express*, vol. 13, no. 4, pp. 1334–1345, 2005.

[94] Y. Okawachi, M. Foster, J. Sharping, A. Gaeta, Q. Xu, and M. Lipson, "All-optical slow-light on a photonic chip," *Opt. Express*, vol. 14, no. 6, pp. 2317–2322, 2006.

[95] F. Morichetti, A. Melloni, A. Breda, A. Canciamilla, C. Ferrari, and M. Martinelli, "A reconfigurable architecture for continuously variable optical slow-wave delay lines," *Opt. Express*, vol. 15, no. 25, pp. 17 273–17 282, 2007.

[96] F. Öhman, K. Yvind, and J. Mørk, "Slow light in a semiconductor waveguide for true-time delay applications in microwave photonics," *IEEE Photon. Technol. Lett.*, vol. 19, no. 15, pp. 1145–1147, 2007.

[97] B. Zhang, L. Yan, L. Zhang, and A. E. Willner, "Multichannel SBS slow light using spectrally sliced incoherent pumping," *J. Lightw. Technol.*, vol. 26, no. 23, pp. 3763–3769, 2008.

[98] Y. Okawachi, J. E. Sharping, C. Xu, and A. L. Gaeta, "Large tunable optical delays via self-phase modulation and dispersion," *Opt. Express*, vol. 14, no. 25, pp. 12 022–12 027, 2006.

[99] A. Yeniay, J.-M. Delavaux, and J. Toulouse, "Spontaneous and stimulated Brillouin scattering gain spectra in optical fibers," *J. Lightw. Technol.*, vol. 20, no. 8, pp. 1425–1432, 2002.

[100] T. F. Krauss, "Why do we need slow light?" *Nature Photonics*, vol. 2, no. 8, pp. 448–450, 2008.

[101] R. W. Boyd, D. J. Gauthier, and A. L. Gaeta, "Applications of slow light in telecommunications," *Opt. Photon. News*, vol. 17, no. 4, pp. 18–23, 2006.

[102] P. Bernasconi, J. E. Simsarian, J. Gripp, M. Dülk, and D. T. Neilson, "Toward optical packet switching," *Photonics Spectra*, pp. 84–93, March 2006.

[103] Y. Liu, E. Tangdiongga, Z. Li, S. Zhang, M. T. Hill, J. H. C. van Zantvoort, F. M. Huijskens, H. de Waardt, M. K. Smit, A. M. J. Koonen, G. D. Khoe, and H. J. S. Dorren, "Ultra-fast all-optical signal processing: toward optical packetswitching," in *Proc. of SPIE*, vol. 6353, no. 1. Gwangju, South Korea: SPIE, Sept. 2006, p. 635312.

[104] L. Zhang, T. Luo, C. Yu, W. Zhang, and A. E. Willner, "Pattern dependence of data distortion in slow-light elements," *J. Lightw. Technol.*, vol. 25, no. 7, pp. 1754–1760, 2007.

[105] C. Yu, T. Luo, L. Zhang, and A. E. Willner, "Data pulse distortion induced by a slow-light tunable delay line in optical fiber," *Opt. Lett.*, vol. 32, no. 1, pp. 20–22, 2007.

[106] B. Zhang, L. Yan, L. Zhang, S. Nuccio, L. Christen, T. Wu, and A. E. Willner, "Spectrally efficient slow light using multilevel phase-modulated formats," *Opt. Lett.*, vol. 33, no. 1, pp. 55–57, 2008.

[107] B. Zhang, L. Yan, I. Fazal, L. Zhang, A. E. Willner, Z. Zhu, and D. J. Gauthier, "Slow light on Gbit/s differential-phase-shift-keying signals," *Opt. Express*, vol. 15, no. 4, pp. 1878–1883, 2007.

[108] L. Yi, Y. Jaouen, W. Hu, Y. Su, and S. Bigo, "Improved slow-light performance of 10 Gb/s NRZ, PSBT and DPSK signals in fiber broadband SBS," *Opt. Express*, vol. 15, no. 25, pp. 16 972–16 979, 2007.

[109] B. Zhang, L. Zhang, L.-S. Yan, I. Fazal, J.-Y. Yang, and A. E. Willner, "Continuously-tunable, bit-rate variable OTDM using broadband SBS slow-light delay line," *Opt. Express*, vol. 15, no. 13, pp. 8317–8322, 2007.

[110] B. Zhang, L.-S. Yan, J.-Y. Yang, I. Fazal, and A. Willner, "A single slow-light element for independent delay control and synchronization on multiple Gb/s data channels," *IEEE Photon. Technol. Lett.*, vol. 19, no. 14, pp. 1081–1083, 2007.

[111] I. Fazal, O. Yilmaz, S. Nuccio, B. Zhang, A. E. Willner, C. Langrock, and M. M. Fejer, "Optical data packet synchronization and multiplexing using a tunable optical delay based on wavelength conversion and inter-channel chromatic dispersion," *Opt. Express*, vol. 15, no. 17, pp. 10 492–10 497, 2007.

[112] Z. Shi, R. W. Boyd, D. J. Gauthier, and C. C. Dudley, "Enhancing the spectral sensitivity of interferometers using slow-light media," *Opt. Lett.*, vol. 32, no. 8, pp. 915–917, 2007.

[113] Z. Shi and R. W. Boyd, "Slow-light interferometry: practical limitations to spectroscopic performance," *J. Opt. Soc. Am. B*, vol. 25, no. 12, pp. C136–C143, 2008.

[114] Z. Shi, R. W. Boyd, R. M. Camacho, P. K. Vudyasetu, and J. C. Howell, "Slow-light Fourier transform interferometer," *Phys. Rev. Lett.*, vol. 99, no. 24, p. 240801, 2007.

[115] E. Mateo, F. Yaman, and G. Li, "Control of four-wave mixing phase-matching condition using the Brillouin slow-light effect in fibers," *Opt. Lett.*, vol. 33, no. 5, pp. 488–490, 2008.

[116] I. Frigyes and A. Seeds, "Optically generated true-time delay in phased-array antennas," *IEEE Trans. Microw. Theory Tech.*, vol. 43, no. 9, pp. 2378–2386, 1995.

[117] Z. Dutton and L. V. Hau, "Storing and processing optical information with ultraslow light in Bose-Einstein condensates," *Phys. Rev. A*, vol. 70, no. 5, p. 053831, 2004.

[118] H. Lee and G. Agrawal, "Suppression of stimulated Brillouin scattering in optical fibers using fiber Bragg gratings," *Opt. Express*, vol. 11, no. 25, pp. 3467–3472, 2003.

[119] P. Weßels, P. Adel, M. Auerbach, D. Wandt, and C. Fallnich, "Novel suppression scheme for Brillouin scattering," *Opt. Express*, vol. 12, no. 19, pp. 4443–4448, 2004.

[120] V. I. Kovalev and R. G. Harrison, "Suppression of stimulated Brillouin scattering in high-power single-frequency fiber amplifiers," *Opt. Lett.*, vol. 31, no. 2, pp. 161–163, 2006.

[121] M. Lorenzen, D. Noordegraaf, C. Nielsen, O. Odgaard, L. Gruner-Nielsen, and K. Rottwitt, "Brillouin suppression in a fiber optical parametric amplifier by combining temperature distribution and phase modulation," in *Conference on Optical Fiber Communication and the National Fiber Optic Engineers Conference (OFC/NFOEC) 2008*, San Diego, CA, USA. OSA, Feb. 2008, paper OML1.

[122] S. Yoo, J. Sahu, and J. Nilsson, "Optimized acoustic refractive index profiles for suppression of stimulated Brillouin scattering in large core fibers," in *Conference on Optical Fiber Communication and the National Fiber Optic Engineers Conference (OFC/NFOEC) 2009*. San Diego, CA, USA: OSA, March 2009, paper JWA5.

[123] JDSU, "Suppression of stimulated Brillouin scattering," JDS Uniphase Corporation, Tech. Rep., 2006, [Online]. Available: http://www.jdsu.com/product-literature/brillouinscattering_tn_cms_ae_0306.pdf [Accessed: Dec. 17, 2009].

[124] M. W. Zmuda, "Stimulated Brillouin scattering effects and suppression techniques in high power fiber amplifiers," Ph.D. dissertation, University of New Mexico, Albuquerque, USA, 2009.

[125] S. P. Singh, R. Gangwar, and N. Singh, "Nonlinear scattering effects in optical fibers," *Progress In Electromagnetics Research*, vol. 74, pp. 379–405, 2007.

[126] S. P. Smith, F. Zarinetchi, and S. Ezekiel, "Narrow-linewidth stimulated Brillouin fiber laser and applications," *Opt. Lett.*, vol. 16, no. 6, pp. 393–395, 1991.

[127] D. Stepanov and G. Cowle, "Properties of Brillouin/erbium fiber lasers," *IEEE J. Sel. Topics Quantum Electron.*, vol. 3, no. 4, pp. 1049–1057, 1997.

[128] G. Cowle, D. Yu, and Y. T. Chieng, "Brillouin/erbium fiber lasers," *J. Lightw. Technol.*, vol. 15, no. 7, pp. 1198–1204, 1997.

[129] J. C. Yong, L. Thévenaz, and B. Y. Kim, "Brillouin fiber laser pumped by a DFB laser diode," *J. Lightw. Technol.*, vol. 21, no. 2, pp. 546–554, 2003.

[130] S. Shahi, S. W. Harun, K. Dimyati, and H. Ahmad, "Brillouin fiber laser with significantly reduced gain medium length operating in L-band region," *Progress In Electromagnetics Research Letters*, vol. 8, pp. 143–149, 2009.

[131] C. Atkins, D. Cotter, D. Smith, and R. Wyatt, "Application of Brillouin amplification in coherent optical transmission," *Electr. Lett.*, vol. 22, no. 10, pp. 556–558, 1986.

[132] N. A. Olsson and J. P. van der Ziel, "Cancellation of fiber loss by semiconductor laser pumped Brillouin amplification at 1.5 μm," *Appl. Phys. Lett.*, vol. 48, no. 20, pp. 1329–1330, 1986.

[133] R. Tkach and A. Chraplyvy, "Fibre Brillouin amplifiers," *Optical and Quantum Electronics*, vol. 21, no. 1, pp. S105–S112, 1989.

[134] M. F. Ferreira, J. F. Rocha, and J. L. Pinto, "Analysis of the gain and noise characteristics of fibre Brillouin amplifiers," *Optical and Quantum Electronics*, vol. 26, no. 1, pp. 35–44, 1994.

[135] M. Junker, M. Ammann, A. Schwarzbacher, J. Klinger, K.-U. Lauterbach, and T. Schneider, "A comparative test of Brillouin amplification and erbium-doped fiber amplification for the generation of millimeter waves with low phase noise properties," *IEEE Trans. Microw. Theory Tech.*, vol. 54, no. 4, pp. 1576–1581, 2006.

[136] X. Bao, D. J. Webb, and D. A. Jackson, "Combined distributed temperature and strain sensor based on Brillouin loss in an optical fiber," *Opt. Lett.*, vol. 19, no. 2, pp. 141–143, 1994.

[137] M. Niklès, L. Thévenaz, and P. A. Robert, "Simple distributed fiber sensor based on Brillouin gain spectrum analysis," *Opt. Lett.*, vol. 21, no. 10, pp. 758–760, 1996.

[138] X. Bao, "Optical fiber sensors based on Brillouin scattering," *Opt. Photon. News*, vol. 20, no. 9, pp. 40–45, 2009.

[139] Y. Dong, X. Bao, and W. Li, "Differential Brillouin gain for improving the temperature accuracy and spatial resolution in a long-distance distributed fiber sensor," *Appl. Opt.*, vol. 48, no. 22, pp. 4297–4301, 2009.

[140] T. Tanemura, Y. Takushima, and K. Kikuchi, "Narrowband optical filter, with a variable transmission spectrum, using stimulated Brillouin scattering in optical fiber," *Opt. Lett.*, vol. 27, no. 17, pp. 1552–1554, 2002.

[141] B. Vidal, M. A. Piqueras, and J. Martí, "Tunable and reconfigurable photonic microwave filter based on stimulated Brillouin scattering," *Opt. Lett.*, vol. 32, no. 1, pp. 23–25, 2007.

[142] A. Zadok, A. Eyal, and M. Tur, "Gigahertz-wide optically reconfigurable filters using stimulated Brillouin scattering," *J. Lightw. Technol.*, vol. 25, no. 8, pp. 2168–2174, 2007.

[143] A. Chraplyvy and R. Tkach, "Narrowband tunable optical filter for channel selection in densely packed WDM systems," *Electr. Lett.*, vol. 22, no. 20, pp. 1084–1085, 1986.

[144] R. Tkach, A. Chraplyvy, R. Derosier, and H. Shang, "Optical demodulation and amplification of FSK signals using AlGaAs lasers," *Electr. Lett.*, vol. 24, no. 5, pp. 260–262, 1988.

[145] R. Tkach, A. Chraplyvy, and R. Derosier, "Performance of a WDM network based on stimulated Brillouin scattering," *IEEE Photon. Technol. Lett.*, vol. 1, no. 5, pp. 111–113, 1989.

[146] J. Domingo, J. Pelayo, F. Villuendas, C. Heras, and E. Pellejer, "Very high resolution optical spectrometry by stimulated Brillouin scattering," *IEEE Photon. Technol. Lett.*, vol. 17, no. 4, pp. 855–857, 2005.

[147] T. Schneider, "Wavelength and line width measurement of optical sources with femtometre resolution," *Electr. Lett.*, vol. 41, no. 22, pp. 1234–1235, 2005.

[148] K.-U. Lauterbach and T. Schneider, "Brillouin scattering in optical fibers for high resolution wavelength and line width measurements," in *Conference on Optical Fiber Communication and the National Fiber Optic Engineers Conference (OFC/NFOEC) 2006*. Anaheim, CA, USA: OSA, March 2006, paper OWI11.

[149] K.-U. Lauterbach, T. Schneider, R. Henker, M. Ammann, and A. Schwarzbacher, "Fast and simple high resolution optical spectrum analyzer," in *Conference on Lasers and Electro-Optics and Quantum Electronics and Laser Science (CLEO/QELS) 2008*. San Jose, CA, USA: OSA, May 2008, paper CMU3.

[150] T. Schneider, M. Junker, and D. Hannover, "Generation of millimetre-wave signals by stimulated Brillouin scattering for radio over fibre systems," *Electr. Lett.*, vol. 40, no. 23, pp. 1500–1502, 2004.

[151] T. Schneider, D. Hannover, and M. Junker, "Investigation of Brillouin scattering in optical fibers for the generation of millimeter waves," *J. Lightw. Technol.*, vol. 24, no. 1, pp. 295–304, 2006.

[152] M. Junker, T. Schneider, M. J. Ammann, A. T. Schwarzbacher, K.-U. Lauterbach, R. Henker, and S. Neidhardt, "32 GHz carrier generation and 200 Mbit/s error free data transmission in a radio over fibre system," in *IET Irish Signals and Systems Conference 2008*, Galway, Ireland, June 2008, pp. 356–360.

[153] M. Junker, T. Schneider, K.-U. Lauterbach, R. Henker, M. J. Ammann, and A. T. Schwarzbacher, "1 Gbit/s Radio Over Fiber downlink at a 32 GHz carrier," in *34th European Conference and Exhibition on Optical Communication (ECOC) 2008*. Brussels, Belgium: IEEE, Sept. 2008, paper Tu.3.F.2.

[154] M. Junker, "Investigation of millimetre wave generation by stimulated Brillouin scattering for radio over fibre applications," Ph.D. dissertation, Dublin Institute of Technology, Dublin, Ireland, 2008.

[155] M. DeMerchant, A. Brown, X. Bao, and T. Bremner, "Structural monitoring by use of a Brillouin distributed sensor," *Appl. Opt.*, vol. 38, no. 13, pp. 2755–2759, 1999.

[156] L. Zou, X. Bao, F. Ravet, and L. Chen, "Distributed Brillouin fiber sensor for detecting pipeline buckling in an energy pipe under internal pressure," *Appl. Opt.*, vol. 45, no. 14, pp. 3372–3377, 2006.

[157] E. Hering, R. Martin, and M. Stohrer, *Physik für Ingenieure*, 10th ed. Berlin: Springer, 2008.

[158] E. Voges and K. Petermann, Eds., *Optische Kommunikationstechnik: Handbuch für Wissenschaft und Industrie*, 1st ed. Berlin: Springer, 2002.

[159] G. Agrawal, *Nonlinear Fiber Optics (Optics and Photonics)*, 3rd ed. San Diego: Academic Press, 2001.

[160] K. Krebber, "Ortsauflösende Lichtleitfaser-Sensorik für Temperatur und Dehnung unter Nutzung der stimulierten Brillouin-Streuung basierend auf der Frequenzbereichsanalyse," Ph.D. dissertation, Ruhr-Universität Bochum, Bochum, 2001.

[161] H. Kuchling, *Taschenbuch der Physik*, 19th ed. Fachbuchverlag Leipzig im Carl Hanser Verlag, 2007.

[162] D. Cotter, "Stimulated Brillouin scattering in monomode optical fiber," *Journal of Optical Communications*, vol. 4, no. 1, pp. 10–19, 1983.

[163] M. Niklès, L. Thévenaz, and P. A. Robert, "Brillouin gain spectrum characterization in single-mode optical fibers," *J. Lightw. Technol.*, vol. 15, no. 10, pp. 1842–1851, 1997.

[164] T. Schneider, "Time delay limits of stimulated-Brillouin-scattering-based slow light systems," *Opt. Lett.*, vol. 33, no. 13, pp. 1398–1400, 2008.

[165] T. Schneider, R. Henker, K.-U. Lauterbach, and M. Junker, "Comparison of delay enhancement mechanisms for SBS-based slow light systems," *Opt. Express*, vol. 15, no. 15, pp. 9606–9613, 2007.

[166] R. Billington, "Measurement methods for stimulated Raman and Brillouin scattering in optical fibres," National Physical Laboratory, Tech. Rep. COEM 31, 1999.

[167] R. G. Smith, "Optical power handling capacity of low loss optical fibers as determined by stimulated Raman and Brillouin scattering," *Appl. Opt.*, vol. 11, no. 11, pp. 2489–2494, 1972.

[168] *Definitions and test methods for statistical and non-linear related attributes of single-mode fibre and cable*, ITU-T Recommendation G.650.2, 2005.

[169] D. Cotter, "Observation of stimulated Brillouin scattering in low-loss silica fibre at 1.3 μm," *Electr. Lett.*, vol. 18, no. 12, pp. 495–496, 1982.

[170] A. B. Ruffin, "Stimulated Brillouin scattering: An overview of measurements, system impairments, and applications," in *Symposium on Optical Fiber Measurements*. NIST, Sept. 2004.

[171] P. Bayvel and P. Radmore, "Solutions of the SBS equations in single mode optical fibres and implications for fibre transmission systems," *Electr. Lett.*, vol. 26, no. 7, pp. 434–436, 1990.

[172] X. Mao, R. Tkach, A. Chraplyvy, R. Jopson, and R. Derosier, "Stimulated Brillouin threshold dependence on fiber type and uniformity," *IEEE Photon. Technol. Lett.*, vol. 4, no. 1, pp. 66–69, 1992.

[173] C. Lee and S. Chi, "Measurement of stimulated-Brillouin-scattering threshold for various types of fibers using Brillouin optical-time-domain reflectometer," *IEEE Photon. Technol. Lett.*, vol. 12, no. 6, pp. 672–674, 2000.

[174] R. Stolen, "Polarization effects in fiber Raman and Brillouin lasers," *IEEE J. Quantum Electron.*, vol. 15, no. 10, pp. 1157–1160, 1979.

[175] M. van Deventer and A. Boot, "Polarization properties of stimulated Brillouin scattering in single-mode fibers," *J. Lightw. Technol.*, vol. 12, no. 4, pp. 585–590, 1994.

[176] C. L. Tang, "Saturation and spectral characteristics of the Stokes emission in the stimulated Brillouin process," *Journal of Applied Physics*, vol. 37, no. 8, pp. 2945–2955, 1966.

[177] T. Schneider, M. Junker, and K.-U. Lauterbach, "Theoretical and experimental investigation of Brillouin scattering for the generation of millimeter waves," *J. Opt. Soc. Am. B*, vol. 23, no. 6, pp. 1012–1019, 2006.

[178] A. Wiatrek, *Numerik einer Zweiseitenband-Verstärkung: Die Erzeugung von Millimeterwellen für Radio over Fiber Systeme*. VDM Verlag, 2009.

[179] A. Wiatrek, T. Schneider, M. Junker, R. Henker, K.-U. Lauterbach, A. T. Schwarzbacher, and M. J. Ammann, "Numerical investigation of Brillouin based double sideband amplification for millimeter-wave generation," in *2008 International Students and Young Scientists Workshop 'Photonics and Microsystems'*, Wroclaw/Szklarska Poreba, Poland, June 2008.

[180] G. S. He and S. H. Liu, *Physics of Nonlinear Optics*. Singapore: World Scientific Publishing Co Pte Ltd, 1999.

[181] M. D. Stenner, M. A. Neifeld, Z. Zhu, A. M. Dawes, and D. J. Gauthier, "Distortion management in slow-light pulse delay," *Opt. Express*, vol. 13, no. 25, pp. 9995–10 002, 2005.

[182] R. Pant, M. D. Stenner, and M. A. Neifeld, "Designing optimal gain profiles for slow-light applications," in *Proc. of SPIE*, vol. 6482, no. 1. San Jose, CA, USA: SPIE, Jan. 2007, p. 64820R.

[183] Z. Zhu, D. Gauthier, Y. Okawachi, J. Sharping, A. Gaeta, R. Boyd, and A. Willner, "Numerical study of all-optical slow-light delays via stimulated Brillouin scattering in an optical fiber," *J. Opt. Soc. Am. B*, vol. 22, no. 11, pp. 2378–2384, 2005.

[184] K. Y. Song, M. González-Herráez, and L. Thévenaz, "Long optically controlled delays in optical fibers," *Opt. Lett.*, vol. 30, no. 14, pp. 1782–1784, 2005.

[185] R. Henker, A. Wiatrek, S. Preussler, T. Schneider, M. J. Ammann, and A. T. Schwarzbacher, "Langsames Licht auf der Basis der stimulierten Brillouin Streuung in photonischen Netzen mit unterschiedlich langen Standard-Einmodenfasern," in *11. ITG-Fachtagung Photonische Netze*, vol. 222, Leipzig, Germany, May 2010, pp. 265–270, paper P14.

[186] L. Xing, L. Zhan, S. Luo, and Y. Xia, "High-power low-noise fiber Brillouin amplifier for tunable slow-light delay buffer," *IEEE J. Quantum Electron.*, vol. 44, no. 12, pp. 1133–1138, 2008.

[187] N. Olsson and J. Van Der Ziel, "Fibre Brillouin amplifier with electronically controlled bandwidth," *Electr. Lett.*, vol. 22, no. 9, pp. 488–490, 1986.

[188] T. Schneider, R. Henker, K.-U. Lauterbach, and M. Junker, "Distortion reduction in slow light systems based on stimulated Brillouin scattering," *Opt. Express*, vol. 16, no. 11, pp. 8280–8285, 2008.

[189] M. Junker and T. Schneider, "Brillouin Bandbreitenerweiterung als variabler Verstärker und Filter für WDM Systeme," in 7th *ITG-Fachtagung Photonische Netze*, vol. 193, Leipzig, Germany, Apr. 2006, pp. 215–217.

[190] M. Junker, T. Schneider, K.-U. Lauterbach, M. Ammann, and A. Schwarzbacher, "Flexible Brillouin bandwidth broadening for an amplification, filtering or millimeter wave generation systems," in *Optical Amplifiers and Their Applications Topical Meeting*. Washington D.C., USA: OSA, June 2006, paper JWB42.

[191] T. Sakamoto, T. Yamamoto, K. Shiraki, and T. Kurashima, "Low distortion slow light in flat Brillouin gain spectrum by using optical frequency comb," *Opt. Express*, vol. 16, no. 11, pp. 8026–8032, 2008.

[192] Z. Lu, Y. Dong, and Q. Li, "Slow light in multi-line Brillouin gain spectrum," *Opt. Express*, vol. 15, no. 4, pp. 1871–1877, 2007.

[193] M. González-Herráez, K. Y. Song, and L. Thévenaz, "Arbitrary-bandwidth Brillouin slow light in optical fibers," *Opt. Express*, vol. 14, no. 4, pp. 1395–1400, 2006.

[194] A. M. Dawes, Z. Zhu, and D. J. Gauthier, "Improving the bandwidth of SBS-based slow-light delay," in *Conference on Lasers and Electro-Optics and Quantum Electronics and Laser Science (CLEO/QELS) 2006*. Long Beach, CA, USA: OSA, May 2006, paper CThW1.

[195] Z. Zhu, A. M. C. Dawes, D. J. Gauthier, L. Zhang, and A. E. Willner, "Broadband SBS slow light in an optical fiber," *J. Lightw. Technol.*, vol. 25, no. 1, pp. 201–206, 2007.

[196] T. Schneider, M. Junker, K.-U. Lauterbach, and R. Henker, "Distortion reduction in cascaded slow light delays," *Electr. Lett.*, vol. 42, no. 19, pp. 1110–1111, 2006.

[197] T. Schneider, M. Junker, and K.-U. Lauterbach, "Potential ultra wide slow-light bandwidth enhancement," *Opt. Express*, vol. 14, no. 23, pp. 11 082–11 087, 2006.

[198] E. Shumakher, N. Orbach, A. Nevet, D. Dahan, and G. Eisenstein, "On the balance between delay, bandwidth and signal distortion in slow light systems based on stimulated Brillouin scattering in optical fibers," *Opt. Express*, vol. 14, no. 13, pp. 5877–5884, 2006.

[199] K. Y. Song and K. Hotate, "25 GHz bandwidth Brillouin slow light in optical fibers," *Opt. Lett.*, vol. 32, no. 3, pp. 217–219, 2007.

[200] T. Schneider, R. Henker, K.-U. Lauterbach, and M. Junker, "Slow und Fast Light in photonischen Netzen - ein Überblick," in 8th ITG-Fachtagung Photonische Netze, vol. 201, Leipzig, Germany, May 2007, pp. 17–24.

[201] R. Henker, A. Wiatrek, K.-U. Lauterbach, M. Junker, T. Schneider, M. J. Ammann, and A. T. Schwarzbacher, "A review of slow- and fast-light based on stimulated Brillouin scattering in future optical communication networks," Communications - Scientific Letters of the University of Zilina, vol. 10, no. 4, pp. 45–52, 2008.

[202] S. Chin, M. González-Herráez, and L. Thévenaz, "Zero-gain slow & fast light propagation in an optical fiber," Opt. Express, vol. 14, no. 22, pp. 10 684–10 692, 2006.

[203] T. Schneider, M. Junker, and K.-U. Lauterbach, "Time delay enhancement in stimulated-Brillouin-scattering-based slow-light systems," Opt. Lett., vol. 32, no. 3, pp. 220–222, 2007.

[204] R. Henker, T. Schneider, M. Junker, K.-U. Lauterbach, M. J. Ammann, and A. T. Schwarzbacher, "Enhancement of maximum time delay in one fiber segment slow light systems based on stimulated Brillouin scattering," in Conference on Lasers and Electro-Optics and Quantum Electronics and Laser Science (CLEO/QELS) 2007. Baltimore, MD, USA: OSA, May 2007, paper CFD4.

[205] T. Schneider, M. Junker, K.-U. Lauterbach, and R. Henker, "Gain-independent SBS based slow light in optical fibers," in Conference on Optical Fiber Communication and the National Fiber Optic Engineers Conference (OFC/NFOEC) 2007. Anaheim, CA, USA: OSA, March 2007, paper JWA14.

[206] Z. Zhu and D. J. Gauthier, "Nearly transparent SBS slow light in an optical fiber," Opt. Express, vol. 14, no. 16, pp. 7238–7245, 2006.

[207] T. Schneider, R. Henker, K.-U. Lauterbach, and M. Junker, "Adapting Brillouin spectrum for slow light delays," Electr. Lett., vol. 43, no. 12, pp. 682–683, 2007.

[208] T. Schneider, R. Henker, K.-U. Lauterbach, and M. Junker, "Adjusting the Brillouin spectrum in optical fibers for slow and fast light applications," in Slow and Fast Light Conference 2007. Salt Lake City, UT, USA: OSA, July 2007, paper SWC3.

[209] R. Henker, A. Wiatrek, K.-U. Lauterbach, T. Schneider, M. J. Ammann, and A. T. Schwarzbacher, "Optimization of the Brillouin spectrum for fiber based slow light systems," in Conference on Lasers and Electro-Optics and Quantum Electronics and Laser Science (CLEO/QELS) 2008. San Jose, CA, USA: OSA, May 2008, paper CThE1.

[210] R. Henker, T. Schneider, A. Wiatrek, K.-U. Lauterbach, M. Junker, M. J. Ammann, and A. T. Schwarzbacher, "Optimisation of optical signal delay in slow-light systems based

on stimulated Brillouin scattering," in *IET Irish Signals and Systems Conference 2008*, Galway, Ireland, June 2008, paper PS-2.5.

[211] T. Schneider, R. Henker, M. Junker, and K.-U. Lauterbach, "Adapting the slow light spectrum in optical fibers for delay enhancement," in *33rd European Conference on Optical Communication (ECOC) 2007*. Berlin, Germany: IEEE, Sept. 2007, paper P010.

[212] R. Henker, K.-U. Lauterbach, A. Wiatrek, T. Schneider, M. J. Ammann, and A. T. Schwarzbacher, "Gain enhancement in slow-light systems based on stimulated Brillouin-scattering with several short fibers," in *Conference on Optical Fiber Communication and the National Fiber Optic Engineers Conference (OFC/NFOEC) 2009*. San Diego, CA, USA: OSA, March 2009, paper OWU5.

[213] R. Henker, A. Wiatrek, S. Preussler, M. J. Ammann, A. T. Schwarzbacher, and T. Schneider, "Gain enhancement in multiple-pump-line Brillouin-based slow light systems by using fiber segments and filter stages," *Appl. Opt.*, vol. 48, no. 29, pp. 5583–5588, 2009.

[214] T. Schneider, A. Wiatrek, and R. Henker, "Zero-broadening and pulse compression slow light in an optical fiber at high pulse delays," *Opt. Express*, vol. 16, no. 20, pp. 15 617–15 622, 2008.

[215] T. Schneider, A. Wiatrek, and R. Henker, "Dispersion compensation by SBS based slow-light in an optical fiber," in *Conference on Optical Fiber Communication and the National Fiber Optic Engineers Conference (OFC/NFOEC) 2009*. San Diego, CA, USA: OSA, March 2009, paper JWA7.

[216] S. Wang, L. Ren, Y. Liu, and Y. Tomita, "Zero-broadening SBS slow light propagation in an optical fiber using two broadband pumpbeams," *Opt. Express*, vol. 16, no. 11, pp. 8067–8076, 2008.

[217] Z. Shi, R. Pant, Z. Zhu, M. D. Stenner, M. A. Neifeld, D. J. Gauthier, and R. W. Boyd, "Design of a tunable time-delay element using multiple gain lines for increased fractional delay with high data fidelity," *Opt. Lett.*, vol. 32, no. 14, pp. 1986–1988, 2007.

[218] R. Pant, M. D. Stenner, M. A. Neifeld, and D. J. Gauthier, "Optimal pump profile designs for broadband SBS slow-light systems," *Opt. Express*, vol. 16, no. 4, pp. 2764–2777, 2008.

[219] A. Zadok, A. Eyal, and M. Tur, "Extended delay of broadband signals in stimulated Brillouin scattering slow light using synthesized pump chirp," *Opt. Express*, vol. 14, no. 19, pp. 8498–8505, 2006.

[220] E. Cabrera-Granado, O. G. Calderón, S. Melle, and D. J. Gauthier, "Observation of large 10-Gb/s SBS slow light delay with low distortion using an optimized gain profile," *Opt. Express*, vol. 16, no. 20, pp. 16 032–16 042, 2008.

[221] A. Wiatrek, R. Henker, S. Preußler, M. J. Ammann, A. T. Schwarzbacher, and T. Schneider, "Zero-broadening measurement in Brillouin based slow-light delays," *Opt. Express*, vol. 17, no. 2, pp. 797–802, 2009.

[222] A. Wiatrek, R. Henker, S. Preußler, and T. Schneider, "Pulse broadening cancellation in cascaded slow-light delays," *Opt. Express*, vol. 17, no. 9, pp. 7586–7591, 2009.

[223] A. Wiatrek, K. Jamshidi, R. Henker, S. Preußler, and T. Schneider, "Nonlinear Brillouin based slow-light system for almost distortion-free pulse delay," *J. Opt. Soc. Am. B*, vol. 27, no. 3, pp. 544–549, 2010.

[224] S. Chin, M. González-Herráez, and L. Thévenaz, "Complete compensation of pulse broadening in an amplifier-based slow light system using a nonlinear regeneration element," *Opt. Express*, vol. 17, no. 24, pp. 21 910–21 917, 2009.

[225] L. Xing, L. Zhan, L. Yi, and Y. Xia, "Storage capacity of slow-light tunable optical buffers based on fiber Brillouin amplifiers for real signal bit streams," *Opt. Express*, vol. 15, no. 16, pp. 10 189–10 195, 2007.

[226] R. Henker, A. Wiatrek, K.-U. Lauterbach, M. J. Ammann, A. T. Schwarzbacher, and T. Schneider, "Group velocity dispersion reduction in fibre-based slow-light systems via stimulated Brillouin scattering," *Electr. Lett.*, vol. 44, no. 20, pp. 1185–1186, 2008.

[227] R. W. Boyd and P. Narum, "Slow- and fast-light: fundamental limitations," *Journal of Modern Optics*, vol. 54, no. 16–17, pp. 2403–2411, 2007.

[228] S. Chin, M. González-Herráez, and L. Thevenaz, "Self-advanced fast light propagation in an optical fiber based on Brillouin scattering," *Opt. Express*, vol. 16, no. 16, pp. 12 181–12 189, 2008.

[229] K. Y. Song, M. González-Herráez, and L. Thévenaz, "Gain-assisted pulse advancement using single and double Brillouin gain peaks in optical fibers," *Opt. Express*, vol. 13, no. 24, pp. 9758–9765, 2005.

[230] S. Chin, M. González-Herráez, and L. Thévenaz, "Simple technique to achieve fast light in gain regime using Brillouin scattering," *Opt. Express*, vol. 15, no. 17, pp. 10 814–10 821, 2007.

[231] R. W. Boyd, D. J. Gauthier, A. L. Gaeta, and A. Willner, "Limits on the time delay induced by slow-light propagation," in *Conference on Lasers and Electro-Optics and Quantum Electronics and Laser Science (CLEO/QELS) 2005*. Baltimore, MD, USA: OSA, May 2005, paper QTuC1.

[232] R. W. Boyd, D. J. Gauthier, A. L. Gaeta, and A. E. Willner, "Maximum time delay achievable on propagation through a slow-light medium," *Phys. Rev. A*, vol. 71, no. 2, p. 023801, 2005.

[233] J. B. Khurgin, "Performance limits of delay lines based on optical amplifiers," *Opt. Lett.*, vol. 31, no. 7, pp. 948–950, 2006.

[234] T. Schneider, R. Henker, K.-U. Lauterbach, and M. Junker, "Delay limits of SBS based slow light," in *Slow and Fast Light Conference 2008*. Boston, MA, USA: OSA, July 2008, paper STuC4.

[235] T. Schneider, R. Henker, A. Wiatrek, K.-U. Lauterbach, and M. Junker, "Grenzen von Slow-Light in photonischen Netzen," in 9th *ITG-Fachtagung Photonische Netze*, vol. 207, Leipzig, Germany, Apr. 2008, pp. 131–136.

[236] C. J. Misas, P. Petropoulos, and D. J. Richardson, "Slowing of pulses to c/10 with subwatt power levels and low latency using Brillouin amplification in a bismuth-oxide optical fiber," *J. Lightw. Technol.*, vol. 25, no. 1, pp. 216–221, 2007.

[237] G. Qin, H. Sotobayashi, M. Tsuchiya, A. Mori, T. Suzuki, and Y. Ohishi, "Stimulated Brillouin scattering in a single-mode tellurite fiber for amplification, lasing, and slow light generation," *J. Lightw. Technol.*, vol. 26, no. 5, pp. 492–498, 2008.

[238] K. S. Abedin, "Stimulated Brillouin scattering in single-mode tellurite glass fiber," *Opt. Express*, vol. 14, no. 24, pp. 11 766–11 772, 2006.

[239] K. Y. Song, K. S. Abedin, and K. Hotate, "Gain-assisted superluminal propagation in tellurite glass fiber based on stimulated Brillouin scattering," *Opt. Express*, vol. 16, no. 1, pp. 225–230, 2008.

[240] K. S. Abedin, "Observation of strong stimulated Brillouin scattering in single-mode As_2Se_3 chalcogenide fiber," *Opt. Express*, vol. 13, no. 25, pp. 10 266–10 271, 2005.

[241] C. Florea, M. Bashkansky, Z. Dutton, J. Sanghera, P. Pureza, and I. Aggarwal, "Stimulated Brillouin scattering in single-mode As_2S_3 and As_2Se_3 chalcogenide fibers," *Opt. Express*, vol. 14, no. 25, pp. 12 063–12 070, 2006.

[242] K. Y. Song, K. S. Abedin, K. Hotate, M. González-Herráez, and L. Thévenaz, "Highly efficient Brillouin slow and fast light using As_2Se_3 chalcogenide fiber," *Opt. Express*, vol. 14, no. 13, pp. 5860–5865, 2006.

[243] M. Lee, R. Pant, and M. A. Neifeld, "Improved slow-light delay performance of a broadband stimulated Brillouin scattering system using fiber Bragg gratings," *Appl. Opt.*, vol. 47, no. 34, pp. 6404–6415, 2008.

[244] C. Jáuregui, P. Petropoulos, and D. J. Richardson, "Brillouin assisted slow-light enhancement via Fabry-Perot cavity effects," *Opt. Express*, vol. 15, no. 8, pp. 5126–5135, 2007.

[245] A. Wiatrek, R. Henker, and T. Schneider, "Almost distortion-free 1.2 bit Brillouin based slow-light," in *34th European Conference on Optical Communication (ECOC) 2009*. Vienna, Austria: IEEE, Sept. 2009, paper 5.1.2.

[246] R. Henker and T. Schneider, "Verfahren und Vorrichtung zur Erzeugung eines Pumpspektrums," German Patent DE 10 2007 058 049 A1, June 04, 2009.

Appendix A

Measurement of Brillouin Parameters

A.1 Brillouin Shift Measurement

Various methods can be used to determine the Brillouin shift. Three measurements are explained here: a direct measurement by changing the frequency shift of two pump waves, a direct measurement of the Brillouin shift with an optical spectrum analyzer (OSA) and the indirect measurement of the Brillouin spectrum by a heterodyne detection.

The experimental setup of the first measurement is shown in Figure A.1a. A laser diode (LD) creates a continuous wave (CW) at a carrier wavelength λ_c around 1550 nm (inset a of Figure A.1a). This CW is externally intensity modulated by a Mach-Zehnder modulator (MZM) with a sinusoidal signal of a frequency f_{mod}. To achieve the maximum output power of the MZM, the polarization of the input wave has to be adapted by an optical polarization controller (OPC). The bias voltage is adjusted to drive the MZM in suppressed carrier regime. Additionally, the input power of the electrical modulation signal is chosen so that only the first-order sidebands are created (inset b of Figure A.1a). The power of these two waves are amplified by an EDFA over the Brillouin threshold. The two pump waves are launched into the fiber under test (FUT) via an optical circulator (C, port $1 \Rightarrow 2$). At the end of the fiber an optical isolator (ISO) prevents spurious backreflections. Inside the FUT each pump wave

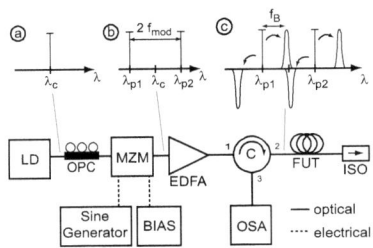

(a) Direct f_B measurement by changing the frequency shift between two pump waves. The insets illustrate the spectra at selected measuring points.

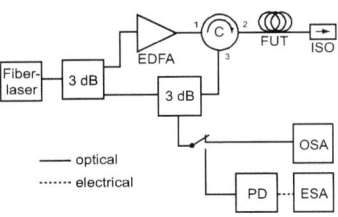

(b) Direct f_B measurement with an OSA and indirect f_B measurement via heterodyne detection.

Figure A.1: Measurement setups for the Brillouin shift.

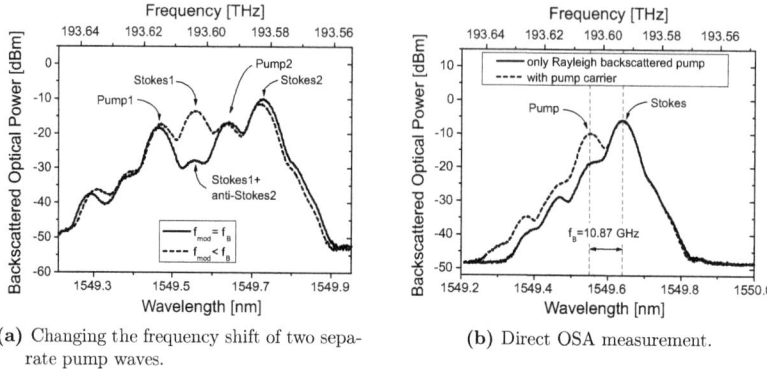

(a) Changing the frequency shift of two separate pump waves.

(b) Direct OSA measurement.

Figure A.2: Output spectra of Brillouin shift measurement.

creates a backscattered Stokes-wave and an anti-Stokes absorption region simultaneously by SBS (inset c of Figure A.1a). If the frequency distance of both pump waves equals two times f_B the Stokes-wave of the first pump wave is canceled out by the anti-Stokes absorption of the second pump wave at λ_c. Therefore, the modulation frequency of the MZM has to be exactly the Brillouin shift. The additional Stokes-wave and the anti-Stokes absorption of the second and first pump waves have no influence on the measurement of f_B and can be neglected. At the output of C (port 3) the spectrum of the backscattered Stokes-waves can be measured by an OSA.

In Figure A.2a the output spectra with a mismatch between f_{mod}, f_B and with $f_{mod} = f_B$ are shown. As can be seen, the backscattered Stokes-wave at λ_c vanishes for $f_{mod} = f_B$. The value of the Brillouin shift can be directly read from the sine wave generator. A frequency of 10.87 GHz was determined for the fibers used in this work.

In Figure A.1b the experimental setup to determine the Brillouin shift via direct measurement with an OSA and heterodyne detection is shown. In this case the CW at λ_c is generated by a fiber laser with a linewidth of $< 1\,\text{kHz}$. The CW is divided into two paths by a 3-dB-coupler. The light in the upper path is used as pump wave and its pump power is amplified over the Brillouin threshold by an EDFA. The pump wave is coupled into the FUT which is terminated with an ISO. Inside the FUT the SBS creates a wavelength upshifted Stokes-wave. At the output of the circulator the backscattered Stokes-wave is combined with the carrier CW by another 3-dB-coupler. The combined signal can either be measured by an OSA or launched into a photodiode (PD)[1] for the heterodyne detection.

The OSA measures the Stokes-wave and the carrier CW at different wavelengths or fre-

---

[1] To enable the measurement of the Brillouin shift around 10 GHz a frequency resolution of $< 10\,\text{GHz}$ for the OSA and a bandwidth of $> 10\,\text{GHz}$ for the PD are necessary. For the measurements in this work an OSA with a resolution of approximately 7.5 GHz and a PD with a bandwidth of 30 GHz were used.

quencies. The difference of the peak frequencies is equal to the Brillouin shift which can be determined by an integrated 'Delta Marker Function'. An example of the f_B measurement via OSA is shown in Figure A.2b.

A heterodyne detection is the superposition or mixing of two waves with different frequencies in a PD. The result is an electrical output signal at an intermediate frequency which corresponds to the frequency difference of the two optical input waves [158]. In the case of the Brillouin shift measurement the Stokes-wave is mixed with the carrier wave. The electrical output wave occurs exactly at a frequency equal to f_B. The whole Brillouin spectrum can be measured by an electrical spectrum analyzer (ESA) which is shown in Figure A.6a.

For the measurement the Rayleigh backscattered part of the pump wave can be also used, as can be seen in Figure A.2b. Therefore, the combination of the Stokes-wave and the CW is not necessary, if the power sensitivity and the dynamic range of the OSA and the PD is sufficient.

A.2 Brillouin Threshold Measurement

The experimental setup to determine the Brillouin threshold is shown in Figure A.3. A LD which is used as pump source creates a CW. The pump power can be varied by an EDFA in constant output power control mode. The pump wave is coupled into the FUT via the circulator. At the output of the fiber (at point B) the transmitted pump power is measured by an optical power meter (OPM). Thereby, the fiber attenuation is neglected. At the output of the circulator (port 3) the backscattered Stokes-power is measured by an OSA. Thereby, only the peak power of the Stokes-wave at $f_p - f_B$ is determined. This has the advantage that any other power at different frequencies, e.g. the Rayleigh backscattered part of the pump wave, does not affect the measurement.

To determine the threshold the pump power is increased starting at values below the threshold up to values where the pump will be depleted. Hence, the transmitted output power and the backscattered Stokes-power can be described as a function of the input pump power at point A of the fiber. As previously described in Subsection 3.1.2, the Brillouin threshold is the pump power value where the Stokes power strongly increases or the transmitted pump power depletes. The transmitted pump power measurement plots of all fibers used in this work are shown in Figure A.4. In the diagrams the output power has been offset with the fiber attenuation which therefore, was neglected. The determined Brillouin threshold values are summarized in Table A.2 of Section A.4 and are in good agreement with the calculated values.

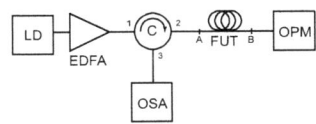

Figure A.3: Measurement setup for Brillouin threshold.

For the measurement of the SSMF with lengths of 0.1 km and 0.2 km the available pump power provided by the equipment was not sufficient to reach the Brillouin threshold.

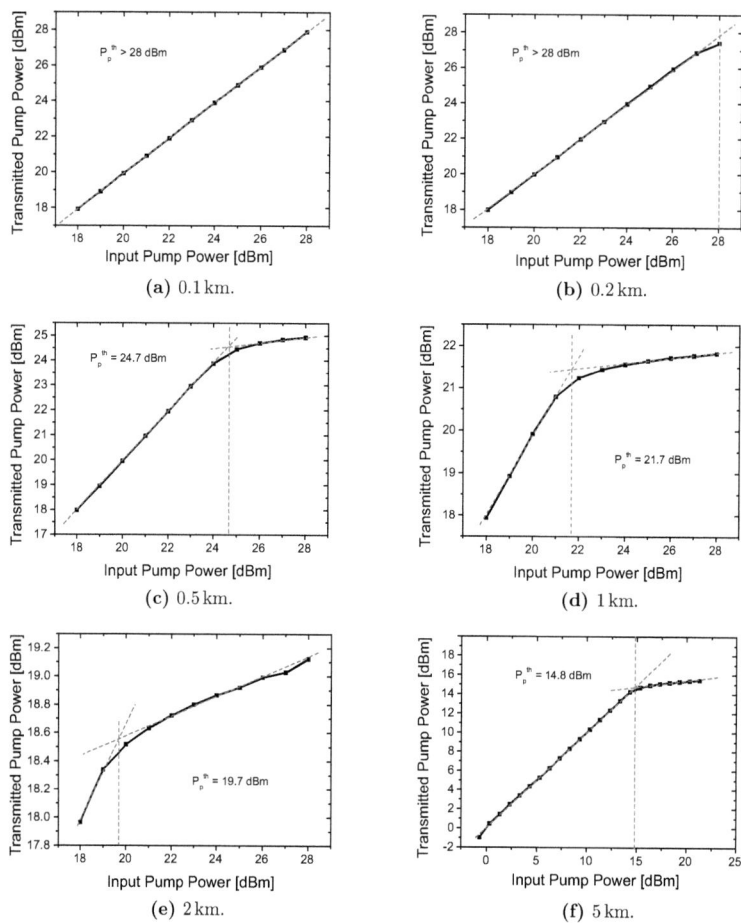

Figure A.4: Brillouin threshold measurement of SSMF with different lengths *(to be continued)*.

A.2 Brillouin Threshold Measurement

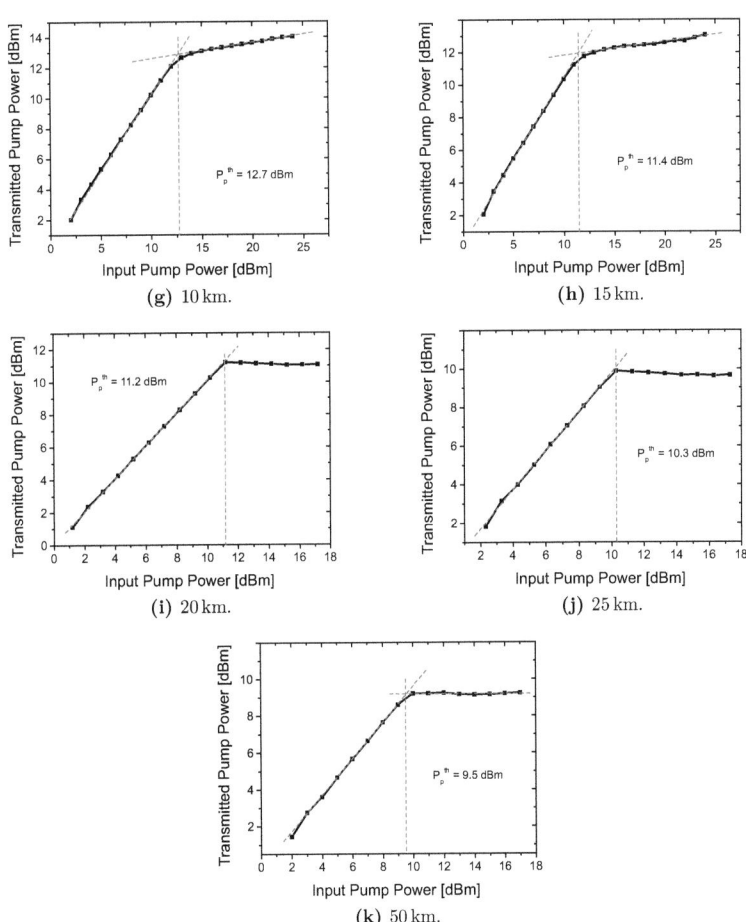

Figure A.4: Brillouin threshold measurement of SSMF with different lengths *(continued)*.

A.3 Brillouin Spectrum and Bandwidth Measurement

One opportunity to measure the Brillouin gain spectrum and the Brillouin bandwidth is the heterodyne detection method described in Section A.1. Therefore, the linewidth of the signal, which is mixed with the Brillouin spectrum in the PD, has to be less than the natural Brillouin bandwidth. The Brillouin spectrum can be measured by an ESA and occurs at the frequency difference of both signals. The measurement result for a 5 km SSMF length is shown in Figure A.6a. As can be seen from the diagram this fiber has a Brillouin shift of 10.87 GHz and a bandwidth of approximately 35 MHz.

Another opportunity to measure the Brillouin spectrum is to use the setup of a high resolution OSA as demonstrated in [147, 148]. This method exploits the narrow bandwidth of the SBS and the property that the resultant Brillouin spectrum is the convolution of the pump and the natural Brillouin spectrum. The modified experimental setup is shown in Figure A.5. The LD generates a carrier CW at λ_c which is divided into two propagation paths (inset a of Figure A.5). In the lower path the carrier is used as pump wave for the SBS generation and is coupled into the FUT from one side via a circulator. In the upper path the carrier is used as signal wave and is launched into the FUT from the opposite side. The isolator prevents the propagation of the pump light back to the LD and protects the LD from damage. Figure A.6b.

An external modulation via MZM produces two sidebands with distance of twice the applied sine frequency. Therefore, the modulator bias voltage and sine amplitude is adjusted to achieve the first-order sidebands and a carrier suppression (inset b of Figure A.5). The maximum output power of the MZM is provided by the OPC. To create a Brillouin gain in the fiber only the lower sideband (in wavelength range) is necessary. Hence, the upper pump wave is filtered out by a tunable filter (TF, inset c of Figure A.5). The EDFA amplifies the pump power to a sufficient high constant value but well below the Brillouin threshold. Inside the FUT a Brillouin gain is created. By varying the modulation frequency of the MZM the Brillouin gain can be crossed over the signal wave which leads to an amplification of the signal wave (inset d of Figure A.5). Thereby, the amplification depends on f_{mod}. If the peak of the Brillouin gain coincides with the peak of the signal wave for instance (at $f_{mod} = f_B$) the maximum output power is achieved. The amplified signal is detected by a PD at the output of the system

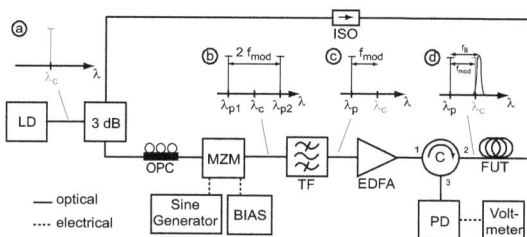

Figure A.5: Measurement setup for Brillouin spectrum. The insets illustrate the spectra at selected measuring points

A.3 Brillouin Spectrum and Bandwidth Measurement

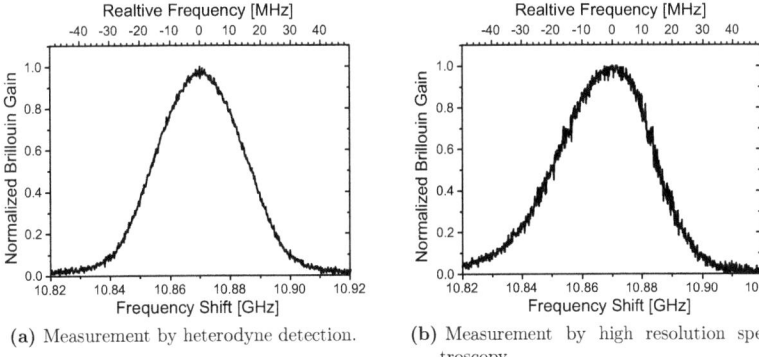

Figure A.6: Brillouin spectrum measurement of a SSMF with a length of 5 km.

(circulator port 3) where the optical power is converted into a DC voltage and measured by a voltmeter. The frequency-dependent output voltage corresponds to the spectral shape of the Brillouin gain, as can be seen in

Although this measurement results in a small distortion, i.e. nonsymmetry, of the Brillouin spectrum, it is similar to the diagram of the heterodyne detection. A Brillouin shift of 10.87 GHz and a bandwidth of approximately 35 MHz can also be read.

A.4 Summary Fiber Parameters

In this section the most important parameters of the used SSMF are summarized. In Table A.1 the values which are equal for every fiber at a reference wavelength of 1550 nm are listed. Thereby, g_{Bmax} and f_B are calculated according Equation (3.6) and Equation (3.2). The Brillouin shift and bandwidth were measured by using the methods described in Section A.1 and Section A.3. The measured g_{Bmax} is an average value which is determined from the measured Brillouin thresholds listed in Table A.2.

Table A.1: Constants of all used fibers.

Parameter	Value	Unit
λ_p	≈ 1550	nm
n	1.44	
v_a	5.96	km/s
α	0.2	dB/km
α	0.046	km^{-1}
M_2	1.51×10^{-15}	s^3/kg
$A_{e\!f\!f}$	86×10^{-12}	m^2
K_B	2	
G_{dB}^{th}	82	dB
f_B (calculated)	≈ 11	GHz
f_B (measured)	10.87	GHz
g_{Bmax} (calculated)	1.92×10^{-11}	m/W
g_{Bmax} (measured)	$\approx 2 \times 10^{-11}$	m/W
Δf_B (measured)	35	MHz

Table A.2 shows the parameters which are different for the used fibers. The effective fiber length and the Brillouin threshold were calculated according Equation (3.5) and Equation (3.4). For the SSMF with lengths of 0.1 km and 0.2 km the exact threshold value could not be measured because the measurement equipment did not provide the necessary pump power. Therefore, these fibers were not considered for the average value of g_{Bmax}. The time delay slopes were determined as described in Subsection 3.2.3 for a natural Brillouin gain. The measured values agree well with the calculated slopes.

A.4 Summary Fiber Parameters

Table A.2: Different parameters of the used fibers.

Parameter	Unit	Value										
L	km	0.1	0.2	0.5	1	2	5	10	15	20	25	50
L_{eff}	km	0.099	0.199	0.494	0.977	1.911	4.466	8.014	10.832	13.070	14.848	19.543
P_p^{th} (calculated)	dBm	32.3	29.3	25.4	22.4	19.5	15.8	13.3	11.9	11.1	10.6	9.4
P_p^{th} (measured)	dBm	>28	>28	24.7	21.7	19.7	14.8	12.7	11.4	11.2	10.3	9.5
g_{Bmax} (measured)	10^{-11} m/W	<5.19	<2.6	1.7	1.75	1.71	2.42	2.19	2.19	1.9	2.05	1.88
$\Delta T/P_p$ (calculated)	ns/mW	<0.14	<0.14	0.26	0.52	1.01	2.36	4.24	5.73	6.91	7.85	10.3
$\Delta T/P_p$ (measured)	ns/mW	0.08	0.14	0.28	0.54	0.95	2.6	4.29	5.8	6.96	8.01	10.5

Appendix B

Pulse Diagrams

In this appendix the normalized time functions of the delayed pulses in comparison to the non-delayed reference pulse are shown. Therefore, the reference pulse or all pulses have been Gaussian-fitted for several selected measurements to enable a better determination of the time delays and pulse widths. The black bold line on the right hand side of the diagrams shows the pulses with the highest measured time delay.

B.1 Single Natural Brillouin Gain

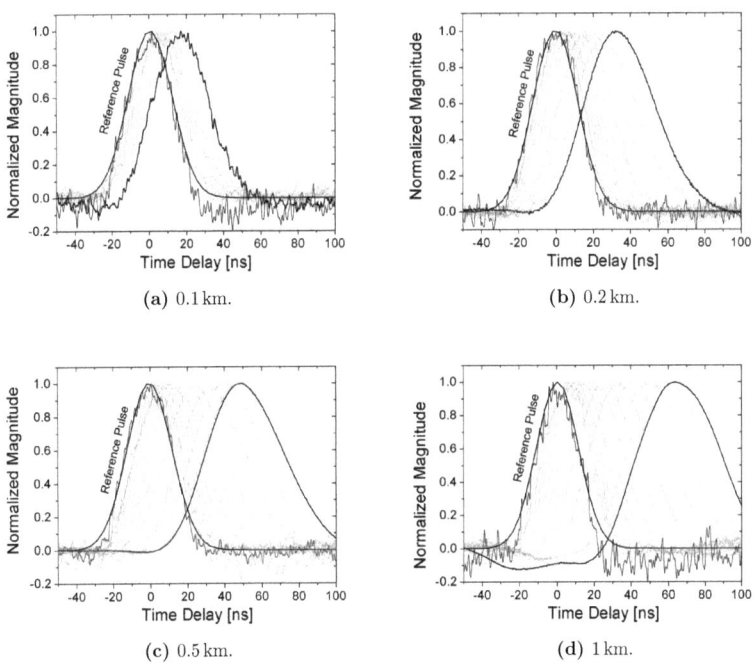

Figure B.1: Delayed pulses for a single natural Brillouin gain *(to be continued)*.

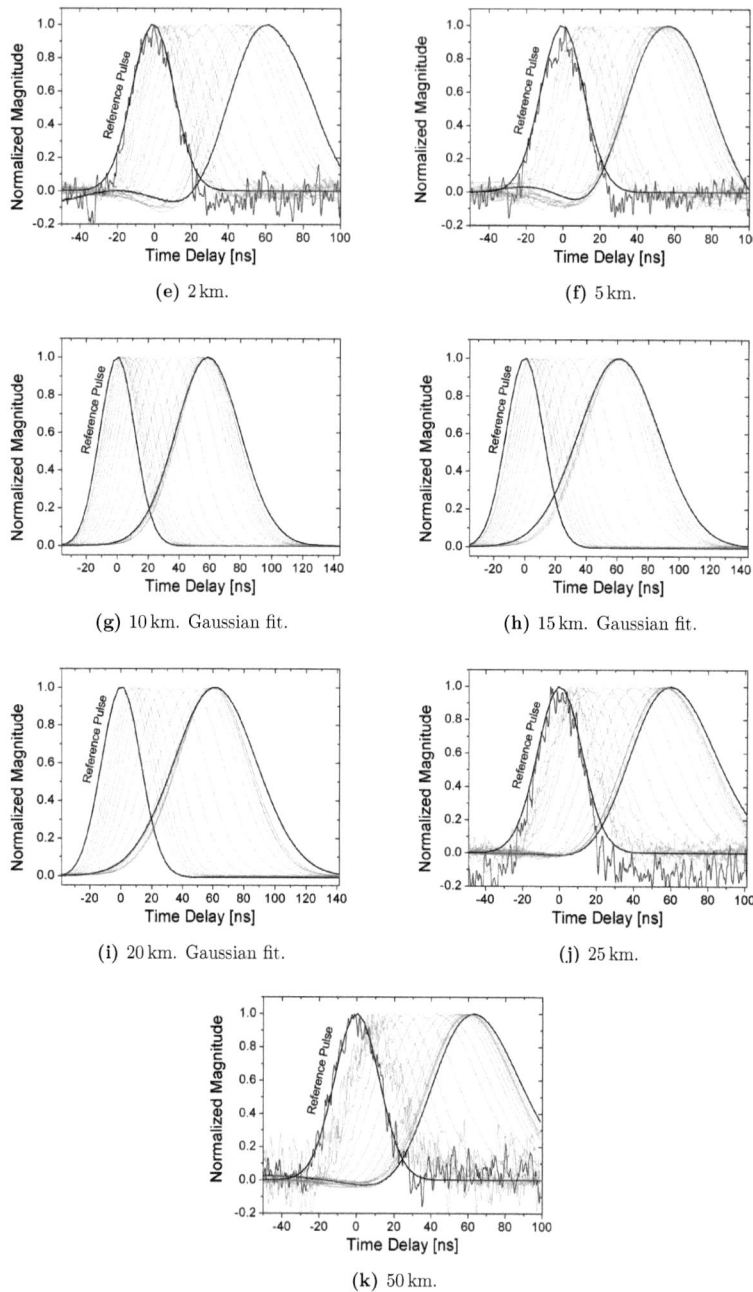

Figure B.1: Delayed pulses for a single natural Brillouin gain *(continued)*.

B.2 Single Narrow Brillouin Gain Superimposed With a Broad Brillouin Loss

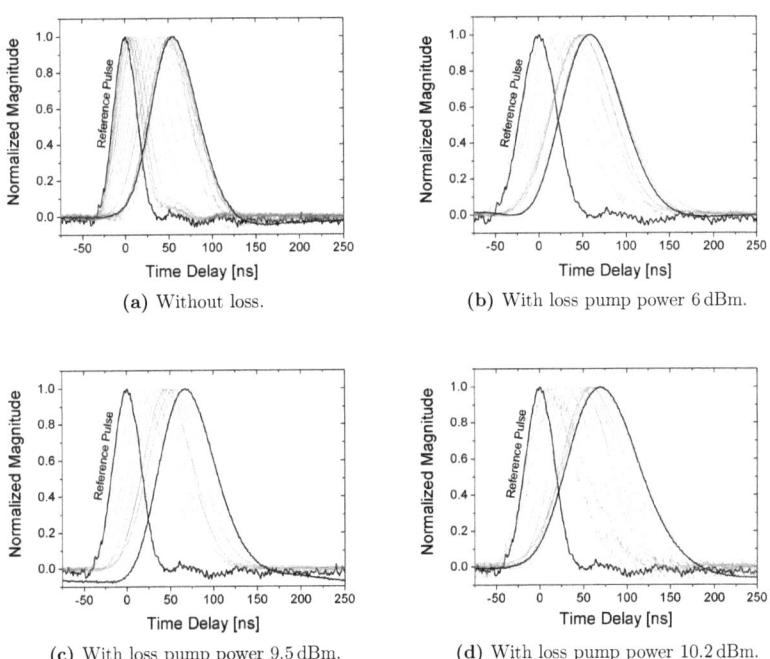

(a) Without loss.

(b) With loss pump power 6 dBm.

(c) With loss pump power 9.5 dBm.

(d) With loss pump power 10.2 dBm.

Figure B.2: Delayed pulses for a superposition of a single natural Brillouin gain with a broad Brillouin loss.

B.3 Single Natural Brillouin Gain Superimposed With Two Narrow Brillouin Losses

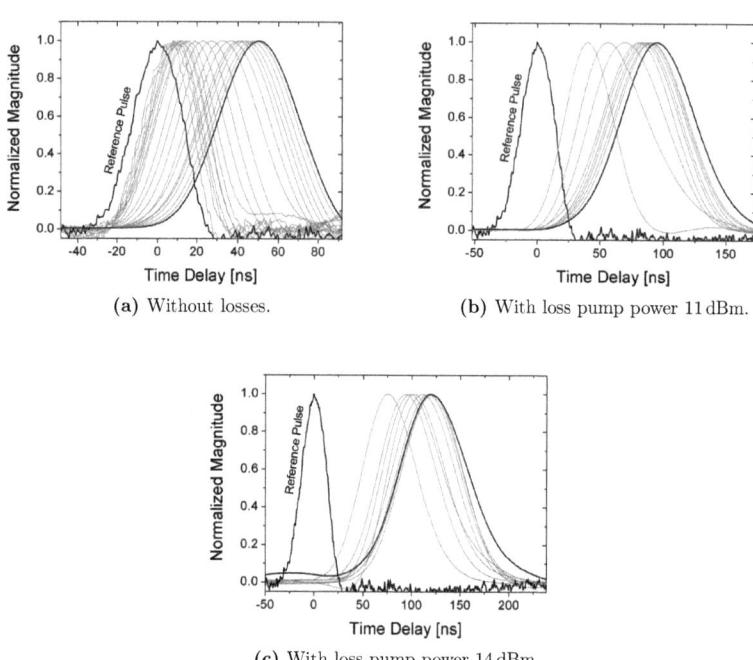

(a) Without losses.

(b) With loss pump power 11 dBm.

(c) With loss pump power 14 dBm.

Figure B.3: Delayed pulses for a superposition of a natural Brillouin gain with two Brillouin losses at its spectral boundaries ($\delta/\gamma_2 = \sqrt{3} \Rightarrow$ loss separation≈ 60 MHz).

B.4 Single Broadened Brillouin Gain Superimposed With Two Narrow Brillouin Losses

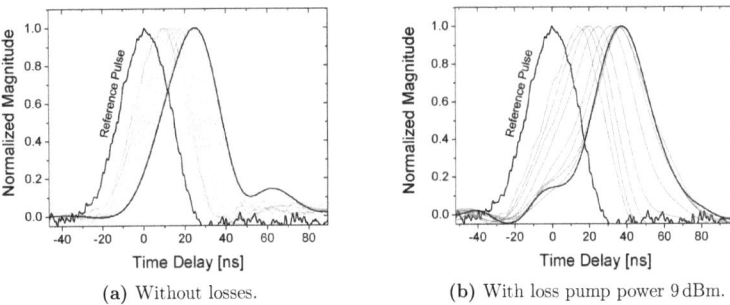

(a) Without losses.

(b) With loss pump power 9 dBm.

Figure B.4: Delayed pulses for a superposition of a broadened Brillouin gain with two Brillouin losses at its spectral boundaries ($\delta/\gamma_2 = \sqrt{3} \Rightarrow$ loss separation$\approx$ 60 MHz, $\Delta f_{B1} \approx$ 60 MHz).

B.5 Suppression of Spurious Backscattered Stokes-Waves via FBG and Filters

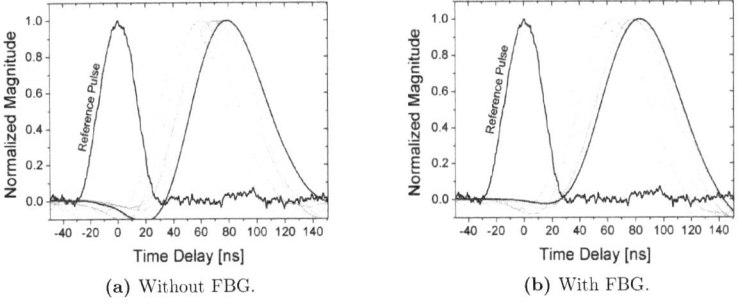

(a) Without FBG.

(b) With FBG.

Figure B.5: Delayed pulses for a blocking via two segments with one FBG.

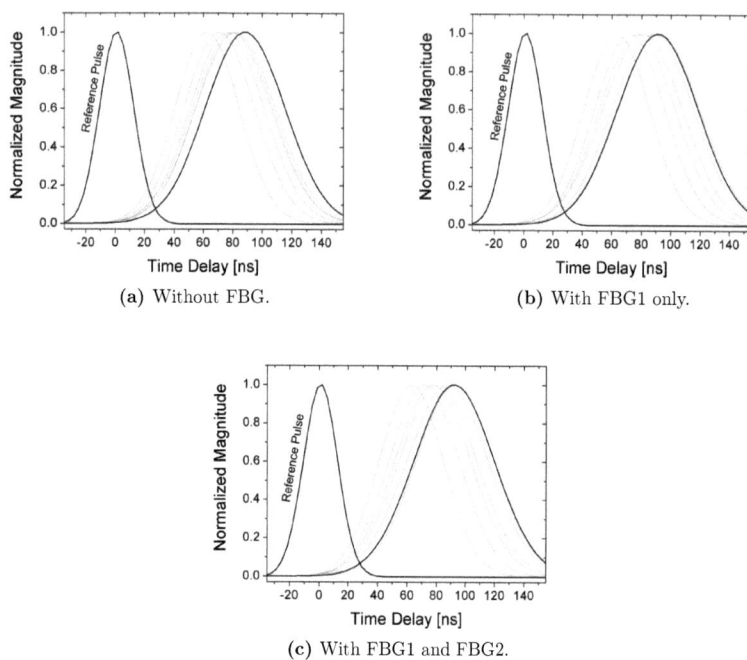

Figure B.6: Delayed pulses for a blocking via three segments with two FBG. Gaussian fit.

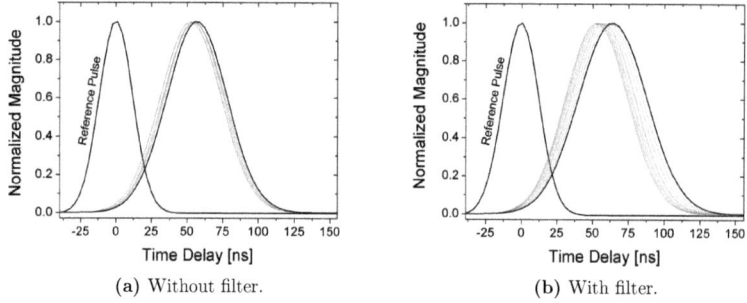

Figure B.7: Delayed pulses for a blocking via two segments with one WDM filter. Gaussian fit.

B.6 Distortion Reduction in Cascaded Slow-Light Systems via External Pump Modulation

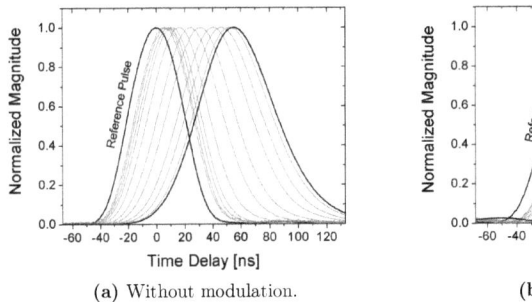
(a) Without modulation.

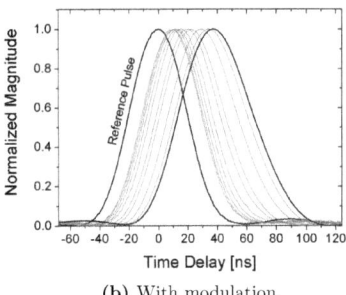
(b) With modulation.

Figure B.8: Delayed pulses for a distortion reduction via external pump modulation in one segment of cascaded slow-light systems.

Index

A

Absorption ... 8, 11, 16, 20, 21, 24, 107–110
- coefficient 11, 12, 20
- resonance 18, 19, 22, 29

All-optical packet router . see Packet router
All-optical packet switch . see Packet switch
Amplified spontaneous emission 22, 59, 114
Amplifier
- erbium-doped fiber 17, 22
- optical 17, 22–23
- semiconductor optical .... see Semiconductor optical amplifier

anti-Stokes-wave .... see Stokes-wave

B

Bit error ratio 15, 29, 59
Bose-Einstein condensate 19, 24, 31
Bragg condition 34
Brillouin
- (frequency) shift 3, 34, 35, 38, 39, 42, 52, 61, 66, 108, 118, 149–151, 154, 156
- amplifier 40, 41, 50, 52, 55, 96
- bandwidth . 2, 3, 38, 41, 42, 48, 50–52, 54–65, 77, 90, 97, 98, 114, 121, 125, 154–156
- gain 3, 4, 35, 40, 41, 43, 46–48, 50, 52, 55, 57, 58, 60, 64, 67, 70, 74, 76, 77, 79, 82, 86, 88, 92, 96, 98, 106–110, 112, 115, 116, 119, 122, 154, 155
- gain coefficient 37–38, 42, 66, 125, 156
- loss 3, 4, 35, 43, 46, 48, 50, 52, 55, 61, 64, 67, 70, 76, 78, 79, 82, 86, 88, 92, 104, 107–112, 114, 116, 119, 122
- spectrum 4, 41, 42, 46, 47, 55, 57, 58, 61, 63–65, 71, 75, 79, 80, 84, 96, 100, 106, 110, 114, 122, 126, 149, 151, 154–155
- threshold 3, 35–38, 40, 42, 65, 66, 87, 92, 107, 109, 115, 116, 149, 151–152, 154, 156
- scattering ... see Stimulated Brillouin scattering

Broadening factor 14, 31, 48, 53, 72, 90, 95–106

C

Coherent population oscillation .. 17, 20–21
Coupled resonator optical waveguide ... see Resonator – CROW

D

Delay
- absolute time 13, 14, 31, 53
- accuracy 15
- effective time 13–14, 31, 53, 55, 65, 77, 83, 86, 95, 97, 106, 113, 114
- fractional time 13–14, 25, 31, 43, 49, 52, 54, 64, 65, 86, 90, 95, 97, 98, 104, 110, 113, 119
- group 8, 13, 18, 25, 46
- negative time 107, 109–111
- phase 8
- pulse 7, 46

– range 15
– resolution 15
Delay-bandwidth product 14–15
Differential phase-shift-keying 29
Dispersion
   – GVD 4, 25, 46, 48, 55, 69, 77, 79,
       99–103, 106, 110, 113
   – anomalous 6, 7, 12, 107–110
   – material 2, 11, 22, 25
   – normal 6, 7, 12, 18, 19
   – relation 6
   – waveguide 18, 21, 25
Doppler effect 34

E
EIT 17–21, 24, 66, 108
Elasto-optic figure of merit 38
Electrostriction 34

F
Fast-light . 3, 5, 7, 12–15, 18, 20–24, 26, 27,
       29–31, 43, 45, 48–50, 54, 55, 57, 64,
       65, 107–113, 123, 125
   – applications 26–31
   – method 13, 16–26
Fiber Bragg grating ... 21, 34, 88–90, 92, 93
Forerunner .... see Precursor
Four-wave mixing . 23, 25, 30, 113, 117, 119

G
Group index 7, 12, 13, 18, 20, 25, 30,
       31, 43–47, 60, 68, 72, 76, 78, 80, 84,
       99–103, 107–109

H
Heterodyne detection 149–151, 154
Hilbert-transform 11

K
Kramers-Kronig relations . 3, 11–12, 19, 23,
       31, 43, 107

O
Optical buffer 27–29, 123

P
Packet router 26–28
Packet switch 26–28
Permeability
   – relative magnetic 11
Permittivity
   – complex relative 11
   – relative 11
Phonon 23, 34, 41
Photonic crystal 17, 22
Polarization 6, 11, 37, 38, 49, 51, 52, 59
Precursor 10
Pseudo-random binary sequence 59
Pump depletion .. 35, 39, 50, 51, 55, 66, 68,
       72, 75, 78, 92, 106, 115

Q
Q-delay product 15
Q-factor 15
Quantum-dot 17, 21, 22
Quantum-well 17, 21–23
Quasi-light storage 17, 26

R
Raman scattering ... see Stimulated Raman
       scattering
Reconfiguration time 15, 25
Refractive index . 6, 7, 11–13, 18, 19, 21, 23,
       47, 60, 68, 75, 78, 99
Resonator
   – CROW 21
   – ring 16, 21, 125

S
Semiconductor
   – (nano)structure 17, 20, 21
   – optical amplifier 17, 22, 23
Silicon-on-insulator structure 21

Slow-light 2, 3, 5, 7, 12–15, 18–24, 26,
    27, 29–31, 43, 45, 48–50, 52–55, 57,
    60, 64, 65, 77, 83, 96, 106, 107, 110,
    112–114, 119, 123, 125
 – applications 26–31
 – method 13, 16–26
Slowdown factor 13
Spontaneous Brillouin scattering 107
Stimulated Brillouin scattering 2, 17, 23, 24,
    31, 33–43, 48, 49, 54, 55, 117
Stimulated Raman scattering 17, 21, 23, 24
Stokes-wave ... 23, 34–41, 66, 86–90, 92, 93,
    106, 108, 115, 150, 151
 – anti 23, 24, 35, 66, 150
Storage capacity . 14, 31, 113, 114, 116, 119,
    123
Susceptibility 11, 117

T

Time delay .... *see* Delay
Transfer function 45

V

Velocity
 – front 9, 10
 – group 1, 3, 6–10, 12, 13, 16,
    18–20, 22–25, 31, 43, 46, 48, 55, 68,
    76, 107, 109, 126
 – information 9–10
 – light in vacuum 5, 6, 10
 – phase 3, 5–7, 13
 – photon 3, 5
 – signal 1, 3, 9–10, 31, 107
 – sound 34, 35
 – superluminal 1, 5, 9, 10, 31

W

Wave number . 6, 45, 46, 55, 67, 74, 77, 108
Waveguide-loops 16, 18, 21, 25
Wavelength-conversion 17, 24, 25, 29

I want morebooks!

Buy your books fast and straightforward online - at one of world's fastest growing online book stores! Environmentally sound due to Print-on-Demand technologies.

Buy your books online at
www.morebooks.shop

Kaufen Sie Ihre Bücher schnell und unkompliziert online – auf einer der am schnellsten wachsenden Buchhandelsplattformen weltweit! Dank Print-On-Demand umwelt- und ressourcenschonend produziert.

Bücher schneller online kaufen
www.morebooks.shop

KS OmniScriptum Publishing
Brivibas gatve 197
LV-1039 Riga, Latvia
Telefax: +371 686 204 55

info@omniscriptum.com
www.omniscriptum.com

Printed by Books on Demand GmbH, Norderstedt / Germany